# Internet Television

# Internet Television

*edited by*

**Eli Noam**
**Jo Groebel**
**Darcy Gerbarg**

Columbia Institute for Tele-Information

THE EUROPEAN INSTITUTE FOR THE MEDIA

Routledge
Taylor & Francis Group
New York   London

First published by:
Lawrence Erlbaum Associates, Inc., Publishers
10 Industrial Avenue
Mahwah, NJ 07430

This edition published 2011 by Routledge:

Routledge
Taylor & Francis Group
711 Third Avenue
New York, NY 10017

Routledge
Taylor & Francis Group
2 Park Square, Milton Park
Abingdon, Oxon OX14 4RN

Cover design by Sean Sciarrone

**Library of Congress Cataloging-in-Publication Data**

Internet television / edited by Eli Noam, Jo Groebel, Darcy Gerbarg.
    p.  cm.
Includes bibliographical references and index.
ISBN 0-8058-4305-1 (c : alk. paper)
ISBN 0-8058-4306-X (pbk. : alk. paper)

TK6679.3 .I59 2003
384.55—dc21                                        2002035400
                                                        CIP

# Contents

# Acknowledgments

This book is the result of a transatlantic collaboration between the Columbia Institute for Tele-Information (CITI) and the European Institute for the Media (EIM). Also participating was the Center for Global Communications at the International University of Japan (GLOCOM). The aim was to look at the advent of widely available individual broadband internet communications and its impact on a new stage in the development of television: Internet television. This global approach produced a broad range of focused, in-depth discussions covering many important issues. The commissioned research papers, collected and edited for this book, provide many insights and much information on Internet television.

This book benefited greatly from the research and administrative help provided by many people. Among them are the following: Reuben Abraham, Keisha E. Burgess, Jason H. Chen, Gabriele Eigen, Raymond Fong, Danilo "Jun" Lopez, Yuko Miyazaki, Rosa M. Morales, Jasmina Pejcinovic, and Stefanie Winde. We thank Robert C. Atkinson, Kenneth R. Carter, Bertram Konert, Koichiro Hayashi, Nobuo Ikeda, and A. Michael Noll for their managerial and substantive contributions to this project. Special thanks go to the authors and to Linda Bathgate, the book's editor at Lawrence Erlbaum Associates.

Michael Einhorn was an especially important collaborator in the project, helping in the conceptualization of the issues and in the identification of leading experts. He deserves much credit.

We are grateful to the Alfred P. Sloan Foundation and its program director, Dr. A. Frank Mayadas, for their support of CITI, as well as to David A. Schaefer of Loeb & Loeb, LLP, and John S. Redpath, Jr., of Home Box Office, Inc. In addition, we wish to thank the State Chancellery North Rhine-Westphalia, Victoria Versicherungen AG, and PriceWaterhouseCoopers for their support of EIM in this project.

Our gratitude is no less to those inadvertently omitted.

*—Eli Noam, Jo Groebel, and Darcy Gerbarg*
New York, February 2003

# Contributor Biographies

**John Carey**
*Managing Director, Greystone Communications, Affiliated Research
Fellow at the Columbia Institute for Tele-Information, and Adjunct
Professor, Columbia University Business School*
John Carey is Managing Director of Greystone Communications, a media
research and planning firm. He conducts research studies of new communication services directed toward homes, businesses, and schools. Currently, he is conducting research about broadband web service,
e-commerce, interactive television, personal video recorders, and digital
satellite radio service for cars.

His clients have included American Express, AT&T, A&E Television Networks, Bell Atlantic, Cablevision, Corporation for Public Broadcasting,
Digitas, Into Networks, Loral Space Systems, NBC, the New York Times Digital Media Company, Public Broadcasting Service, Rogers, Cablesystems,
and XM Satellite Radio, among others.

Dr. Carey is also an Adjunct Professor at Columbia University Business
School, where he teaches graduate courses on Demand for New Media.
He is an Affiliated Research Fellow at the Columbia Institute for Tele-Information. He holds a PhD from the Annenberg School for Communications
at the University of Pennsylvania and is the author of more than 50 publications on interactive media and the adoption of new telecommunication
technologies.

**Kenneth R. Carter**
*Deputy Director, Columbia Institute for Tele-Information, Columbia University Business School*
Kenneth Carter is the Deputy Director of CITI and a candidate for an Executive MBA at Columbia University. He joined the institute in June 1998 as Associate Director. Previously, Mr. Carter worked for the Federal Trade Commission (FTC) on such issues as the FTC's jurisdiction over resellers of prepaid telecommunication services for deceptive advertising of tariff rates. Mr. Carter has a background in media and communications, having worked for MTV Networks, Island Records, and the international television syndication firm D.L. Taffner. As Deputy Director, he manages CITI's research agenda, assists the development of the institute's online research platform, the Virtual Institute of Information, and serves as the institute's counsel. Mr. Carter's current research includes the Emerging Market Economy in Bandwidth, Over the Internet, and the regulatory and intellectual property issues in telecommunications. He received his JD from the Benjamin N. Cardozo School of Law, where he was a member of *The Cardozo Arts & Entertainment Law Journal* and President of the Asian and Pacific Law Students Association. Mr. Carter was awarded an Alexander Judicial Fellowship, serving as a full-time junior clerk in the chambers of Hon. John C. Lifland, U.S.D.J. He was graduated from Colgate University with an A.B. in Economics and East Asian Studies after studying abroad in England and Japan, and is proficient in Japanese. He is presently admitted to the bar in New York State and the District of Columbia.

**Gali Einav**
*PhD candidate, School of Journalism, Researcher at the Interactive Design Lab, Columbia University*
Gali Einav has a BA in political science and an MA in communications and journalism from Hebrew University, Jerusalem. She has worked both as a senior producer and journalist for the second television channel in Israel. Ms. Einav taught media studies at the New School of Communications in Tel Aviv. She is currently a PhD candidate in the communications program and a researcher at the Interactive Design Lab at Columbia University's School of Journalism. Her research interests include content models for interactive media.

**Michael A. Einhorn**
*Principal in the New York office of LECG, LLC*
Michael A. Einhorn is a consultant and testifying expert active in the areas of intellectual property, antitrust, media, and entertainment. Dr. Einhorn is also an Adjunct Professor in the Graduate School of Business at Fordham University, where he teaches a course in the entertainment industry.

Professor Einhorn has considerable professional experience in media and entertainment, having written articles, prepared affidavits, or testified in matters related to music licensing, antitrust, Internet television, peer-to-peer file sharing, digital rights management, anticircumvention, and misuse of copyright. He received a BA from Dartmouth College and a PhD in economics from Yale University. He also served as a professor of economics at Rutgers University and worked as an economist in the Antitrust Division of the U.S. Department of Justice and at Broadcast Music Inc.

## Darcy Gerbarg
*Executive Director, Marconi International Fellowship Foundation and Senior Fellow, Columbia Institute for Tele-Information, Columbia University Business School*
Darcy Gerbarg is a Senior Fellow at the Columbia Institute for Tele-Information, Columbia University Business School, since 1997. In 2000, she was Director of Business Development at Everest Broadband Networks. Prior to that time, she held research positions at Courant Institute for Mathematical Sciences, New York University, and the Computer Graphics Lab, New York Institute of Technology. She has been an adjunct faculty member at the Interactive Telecommunications Graduate Program and the Film and Television Departments at New York University. She was an adjunct faculty member at the State University of New York at Stony Brook, where she also built networked multimedia labs. Ms. Gerbarg started and directed the Graduate Program in Computer Art and the Computer Institute for Arts at the School of Visual Arts in New York. She also initiated and chaired the first SIGGRAPH Computer Art Shows.

Ms. Gerbarg has lectured, organized, and conducted panels, workshops, and presentations at professional conferences for industries, companies, and universities. She has a continuing interest in entrepreneurial activities, start-ups, and venture capital. Conferences she has organized for CITI include the Future of Digital TV (1997) and Venture Capital in New Media (1999). Ms. Gerbarg's edited publications include *The Economics, Technology and Content of Digital TV* (Kluwer Academic Publishers, 1999) and *Digital TV* (*Prometheus*, the Journal of Issues in Technological Change, Innovation, Information Economics, Communications and Science Policy, Carfax Publishing Ltd., June 1998). Ms. Gerbarg has a BA from the University of Pennsylvania and an MBA from New York University.

## Jo Groebel
*Professor and Director, European Institute for the Media, EIM*
Jo Groebel is Director-General of the European Institute for the Media in Düsseldorf and Paris, and holds a professorship for media at the University of Amsterdam. He is a visiting professor at the University of California

in Los Angeles (UCLA) and the University St. Gallen. He was President of the Dutch Association for Communication Sciences, and advisor to the Dutch and German governments at the highest levels, the UN, and several broadcasters and media firms. He is author/editor of 20 books. He has worked on numerous TV and radio productions internationally and contributed to many publications.

## Jeffrey Hart
*Professor of Political Science, Indiana University*
Jeffrey Hart is Professor of Political Science at Indiana University, Bloomington, where he has taught international politics and international political economy since 1981. His first teaching position was at Princeton University from 1973 to 1980. He was a professional staff member of the President's Commission for a National Agenda for the Eighties from 1980 to 1981.

Professor Hart worked at the Office of Technology Assessment of the U.S. Congress in 1985–1986 as an internal contractor and helped to write their report *International Competition in Services* (1987). He was visiting scholar at the Berkeley Roundtable on the International Economy, 1987–1989. His publications include *The New International Economic Order* (1983), *Interdependence in the Post Multilateral Era* (1985), *Rival Capitalists* (1992), and (with Joan Spero) *The Politics of International*.

## Michael L. Katz
*Professor of Business Administration and Economics, Director*
*of the Center for Telecommunications and Digital Convergence,*
*University of California at Berkeley*
Michael L. Katz is the Edward J. and Mollie Arnold Professor of Business Administration at the University of California at Berkeley. He also holds an appointment as Professor in the Department of Economics. Professor Katz is the faculty leader of the Haas Business School's e-business initiatives, and serves as Director of the Center for Telecommunications and Digital Convergence. He is a four-time finalist for the Earl F. Cheit award for outstanding teaching and has won it twice.

Dr. Katz has published numerous articles on the economics of networks industries, intellectual property licensing, telecommunications policy, and cooperative research and development. He is coeditor of the *California Management Review* and serves on the editorial board of the *Journal of Economics and Management Strategy*. Dr. Katz also serves on the Computer Science and Telecommunications Board of the National Academy of Sciences.

Dr. Katz served as Chief Economist of the Federal Communications Commission from January 1994 through January 1996. He participated in the formulation and analysis of policies toward all industries under commission jurisdiction, including broadcasting, cable, telephone, and wireless communications.

Dr. Katz holds an AB *summa cum laude* from Harvard University and a DPhil from Oxford University. Both degrees are in economics.

### Bertram Konert
*Head, Digital World Program the European Institute for the Media,*
*Lecturer, University of Düsseldorf*
Bertram Konert is head of the "Digital World Program" at the European Institute for the Media and lectured in media science at the University of Düsseldorf. After training as a banker, he studied social science at the University of Osnabrück (1982–1987). He started his career as a researcher of telecommunications policy and electronic banking and received his doctorate in the area of economic and social science from the University of Osnabrück in1993. Afterward, he worked for several years as a project manager for a computer company, where he was responsible for customer relations in the area of ISDN-networks and data communications.

Since June 1996, his main research interests at the European Institute for the Media include the socioeconomic developments of media transformation and convergence, particularly in the areas of digital broadcasting and new Internet services. In 2001, Dr. Konert became an editorial advisor on the editorial board of the research journal *Convergence,* published by the University of Luton Press.

### Christopher T. Marsden
*Consultant with Re: Think, www.re-think.com, Marsden@re-think.com,*
*and Research Associate of the Phoenix Center, Washington, DC*
Christopher T. Marsden has wide-ranging experience in academia, the Internet, telecommunications business, and public policy. He was previously Research Fellow (1999–2000) at the Harvard Information Infrastructure Project, Lecturer at Warwick Law School (1997–2000), and LL.M. Supervisor at the London School of Economics (1994–1997). He directed the ESRC European Media Regulation Seminar Group in 1998–1999. He has edited the following books: *Convergence in European Digital TV Regulation* (London: Blackstone, 1999, with Stefaan Verhulst) and *Regulating the Global Information Society* (Routledge, 2000). His current research is in legal, business, and technical challenges to video over Internet protocol, and especially standard setting, which is examined in "Cyberlaw and International Political Economy: Towards Regulation of the Global Information Society" 2001 L.REV. M.S.U.-D.C.L. 1. He contributes to journals including *info, Communications Week International,* and *Inside Digital TV.* In 1998, Mr. Marsden founded the International Journal of Communications Law and Policy (www.ijclp.org), which he coedits. He is also a consultant with London-based digital communications boutique consultancy Re: Think!

(www.re-think.com), and has been expert consultant to the Chief Executive of the Independent Television Commission (2000), the Council of Europe MM-S-PL Committee on digital media pluralism (1999), and Shell International's Global Scenario Planning team (2000). He was UK Regulatory Director of MCI Worldcom from 2001–2002 when he resigned.

**Eli M. Noam**
*Professor of Economics and Finance, and Director of the Columbia*
*Institute for Tele-Information, Columbia University Business School*
Eli Noam is the Professor of Economics and Finance at the Columbia University Business School since 1976. After having served for 3 years as Commissioner of the New York State Public Service Commission, he returned to Columbia in 1990. He served as Director of the Columbia Institute for Tele-Information, an independent university-based research center focusing on strategy, management, and policy issues in telecommunications, computing, and electronic mass media; and Chairman of MBA concentration in the Management of Entertainment, Communications, and Media at the Business School. He has also taught at Columbia Law School and Princeton University's Economics Department and Woodrow Wilson School. Professor Noam has published over 20 books and 400 articles in economic journals, law reviews, and interdisciplinary journals and has served on the editorial boards of other Columbia University Press academic journals. He was a member of the advisory boards for the federal government's FTS-2000 telecommunications network, the IRS's computer system reorganization, and the National Computer Systems Laboratory. He received an AB (Phi Beta Kappa), MA, PhD (Economics) and JD from Harvard University.

**A. Michael Noll**
*Professor of Communications at the University of Southern California,*
*Annenberg School for Communication, Director of Technology*
*Research at the Columbia Institute for Tele-Information, Columbia*
*University Business School*
A. Michael Noll is a Professor of Communications at the Annenberg School for Communication at the University of Southern California. He currently serves as Director of Technology Research at CITI. Professor Noll has had a varied career, including basic research at Bell Labs, science policy on the staff of the White House Science Advisor, and marketing at AT&T. He is an early pioneer in computer art, stereoscopic computer animation, and force-feedback (a forerunner of today's virtual reality). He has published over 75 papers on his research and is the author of seven books on telecommunication science and technology. His current research is focused broadly on the multidisciplinary technological, economic, consumer, business, and policy aspects of telecommunication. Professor Noll is a

seasoned author of op-ed pieces and a frequent columnist in trade magazines. He received his PhD in Electrical Engineering from the Polytechnic Institute of Brooklyn in 1971, MEE from New York University in 1963, and BSEE from Newark College of Engineering in 1961. He is a Senior Member of the Institute of Electrical and Electronics Engineers (IEEE) and is a member of the Audio Engineering Society, the Society for Information Display, and the Society of Motion Picture and Television Engineers.

## Andrew Odlyzko
*Professor of Mathematics, Director Digital Technology Center,*
*Assistant Vice President for Research, University of Minnesota*
Andrew Odlyzko has recently assumed the positions of Professor of Mathematics, Director Digital Technology Center, and Assistant Vice President for Research at the University of Minnesota. Until this year, he was the head of the Mathematics and Cryptography Research Department at AT&T Labs. He has done extensive research in technical areas such as computational complexity, cryptography, number theory, combinatorics, coding theory, analysis, and probability theory. In recent years, he has also been working on electronic publishing, electronic commerce, and economics of data networks. Professor Odlyzko is the author of such widely cited papers as "Tragic Loss or Good Riddance? The Impending Demise of Traditional Scholarly Journals," "The Decline of Unfettered Research," and "The Bumpy Road of Electronic Commerce."

## Robert Pepper
*Chief, Office of Plans and Policy, Federal Communications Commission*
Robert Pepper has been Chief of the Office of Plans and Policy (OPP) at the Federal Communications Commission (FCC) since December 1989. Under Pepper's leadership, OPP is responsible for policy questions that cut across traditional industry and institutional boundaries, especially those arising from the development of new technologies. At OPP, Dr. Pepper's responsibilities have included leading teams implementing provisions of the Telecommunications Act of 1996; assessing the development of the Internet; designing and implementing the first spectrum auctions in the United States; developing more market-based spectrum policies; assessing competition in the video marketplace; and assessing the impact of the development of the Internet on traditional communications policy structures.

Before joining the FCC, Dr. Pepper was Director of the Annenberg Washington Program in Communications Policy Studies. He also has been Director of Domestic Policies and Acting Associate Administrator at the National Telecommunications and Information Administration and developed a program on communications, computers, and information at the National Science Foundation.

He is a graduate of the University of Wisconsin-Madison, where he also received his doctorate.

## Fritz F. Pleitgen
*Director-General, WDR*
Fritz Pleitgen is the Director-General of WDR, Westdeutscher Rundfunk, the largest broadcasting corporation in the German Association of Public Broadcasting Corporations, ARD. He took up this post in 1995. In January 2001, he became chairman of the ARD.

Initially a newspaper journalist, Mr. Pleitgen joined the WDR in 1963 as a reporter for ARD's main news program. In 1970, he was appointed ARD-correspondent in Moscow. He became Head of the ARD Studio in East Berlin in 1977. In 1982, he and his family moved to the United States, where he took over the ARD Studio in Washington. Mr. Pleitgen held this post for 5 years and then became head of the ARD Studio in New York. He returned to the WDR headquarters in Cologne in 1988 to become Editor-in-Chief of WDR television and head of the politics and current affairs section. During this period, Mr. Pleitgen won great acclaim for his reports on German reunification and the collapse of the Soviet Union. He was appointed Radio Director in 1994. In addition, Mr. Pleitgen regularly appears in television programs on WDR and ARD, both as presenter and reporter.

## David Waterman
*Professor, Department of Telecommunications, Indiana University*
David Waterman is Associate Professor in the Department of Telecommunications at Indiana University, Bloomington, since 1993. He was previously a faculty member of the Annenberg School for Communication at the University of Southern California. At USC, Professor Waterman taught in the Annenberg School's Communications Management Masters program and in the Department of Economics. Prior to joining USC, Professor Waterman was the principal of Waterman & Associates, a Los Angeles consulting firm providing economic, policy, and market research services to communications industry and federal government clients. He has also served as Research Economist at the National Endowment for the Arts in Washington.

Professor Waterman has written widely on the economics of the cable television, motion picture, and other information industries. He is coauthor of *Vertical Integration in Cable Television* (MIT Press, 1997) with Andrew A. Weiss. His articles on market structure and public policy toward the media, the economics of motion picture production and distribution, international trade in motion pictures and video products, and other topics have appeared in *Information Economics and Policy, Journal of Communication, Journal of Econometrics, Telecommunications Policy, Federal Communications Law Journal,* and other academic journals and edited books. The Corporation for Public Broadcasting and the National Endowment for the Arts have supported his research. Professor Waterman has presented his research in testimony before the U.S. Congress, and has served on expert panels or in an advisory capacity for the Federal Communications Commission, the Federal Trade Commission, the U.S. Department of Justice, and the General Accounting Office of the United States.

Professor Waterman received a PhD in Economics in 1979 from Stanford University. He completed his BA in Economics at USC.

### Stephen Whittle
*Controller, BBC, Director of the Broadcasting Standards Commission*
Stephen Whittle returned to the BBC as Controller, Editorial Policy, in July 2001. Since 1996, he had held the post of Director of the Broadcasting Standards Commission. Before moving to the BBC, between 1993 and 1996, he was the Chief Adviser, Editorial Policy, of the BBC, and was responsible, among other things, for writing the first edition of the BBC Producer Guidelines. Between 1989 and 1993, he was Head of Religious Programs for the BBC. In 1977, from the World Council of Churches, where he was the Deputy Director of Communications, Mr. Whittle joined the BBC in Manchester.

# Introduction

Darcy Gerbarg and Eli Noam
*Columbia Institute for Tele-Information*

Internet television is the quintessential digital convergence medium, putting together television, telecommunications, the Internet, computer applications, games, and more. It is part of a historic move from individualized narrowband capacity, measured by kilobits per user, to one of broadband with a capacity of megabits per user. This move will have major consequences for many aspects of society and the economy, similar to the impact the automobile had when it replaced trains, horses, and bicycles. It will affect, in particular, the medium now called television.

What exactly is Internet television (TV)? There is no agreement on a definition. It comes with different names—web TV, IPTV, enhanced TV, personal TV, and interactive TV, for example—which signify slightly different things. At the lower end of complexity, it is merely a narrowband two-way Internet-style individualized ("asynchronous") channel that accompanies regular one-way "synchronous" broadband broadcast TV or cable. This Internet channel can provide information in conjunction with broadcast programs, such as details on news and sports, or enable transactions (including e-commerce) in response to TV advertisements. This is known as "enhanced TV." At the other end of complexity is a fully asynchronous two-way TV, with each user receiving and transmitting individualized TV programs, including direct interaction in the program plot line. In between is one-way broadband with a narrowband return channel that can be used to select video programs on demand (VOD). What Internet TV is today and can be in the future forms the context of this book.

This new medium is knocking at the door. Already, music is reaching millions of listeners around the world through the Internet. Video clips have traveled likewise. It will not be long before popular video programs are regularly delivered over the Internet as well, at significantly better quality and lower cost. People with broadband connections already download feature-length films, and in Japan, Yahoo BB is launching a portal of video channels.

Every new medium starts as a substitute and then evolves into something quite new. Internet TV, too, will first be used to access video servers that store existing programs, making them available for viewing at any time. But soon, going beyond the convenience of viewer choice and control, Internet TV will enable and encourage new types of entertainment, education, and games that take advantage of the Internet's interactive capabilities. This assumes, of course, technical capability and economic viability, subjects of analysis in this volume.

This book is organized into five major sections: Infrastructure Implications, Network Business Models and Strategies, Content and Culture, Policy, and Global Impacts. Each section is introduced here.

## INFRASTRUCTURE IMPLICATIONS

Ubiquitous and affordable broadband would enable Internet TV to rival traditional broadcast cable and satellite distribution. However, Internet TV, if used by millions in an asynchronous fashion, would require prodigious amounts of bandwidth. This raises questions about the required network capacity for various quality grades. Delivering individualized broadband to the home is hence a costly and difficult roadblock.

Intermediate solutions include the squeezing of more transmission capacity out of an existing basic infrastructure. Advanced variants of digital subscriber loops (DSL) are one example. Signal compression, caching, and mirroring are other approaches. The infrastructure consists of networks, servers, home terminals, various forms of software, and content. Many approaches require the technical interoperability of a variety of hardware devices and software. Internet TV therefore requires cooperation between several industries, which has contributed to the difficulties in defining technical standards.

This section begins with an introduction to the possibilities of Internet TV by A. Michael Noll of the Annenberg School for Communication at the University of Southern California and the Columbia Institute for Tele-Information. He describes the technologies that make Internet TV possible. Noll reviews the past efforts to introduce broadband video and points to today's technical, business, and other challenges. Noll defines the future of Internet TV as the convergence of broadcast and the Internet. He ends this chapter with a discussion of various possible scenarios and identifies uncertainties in this path.

Andrew Odlyzko, of the University of Minnesota, discusses the implications of Internet TV for long distance networks. He explains why there will be plenty of bandwidth in the backbone of telecom networks to carry video. But, because transmission capacity will be vastly more expensive than storage capacity, the trend will be toward store-and-replay models.

In his second chapter, Noll provides technical specifications and requirements for Internet TV, including reviews of delivery infrastructure and compression. He challenges some technical assumptions and usage data while pointing to problems and solutions arrived at in radio, satellite, and cable distribution. He looks at the changes radio has undergone with the advent of the Internet and projects similar changes for television. He presents various delivery options and reviews alternative technological convergence scenarios, concluding with a review of open technical and other issues that require resolution.

## NETWORK BUSINESS MODELS AND STRATEGIES

Although the technical issues that must be overcome are complex, they pale before the business challenges. Simply put, no one has found a way to make Internet TV a financial success. Early Internet TV content companies struggled to develop a customer base. Some companies tried to create original content to gain new audiences, but this proved very expensive. All seemed to have overestimated the attractiveness of the medium at the time. A major problem encountered was that audiences expect a similar production quality from Internet TV that exists in broadcast and cable television.

The revenue side is equally daunting. Audiences accustomed to receiving Internet and broadcast TV content for free expect Internet TV to be similarly priced. One major challenge for Internet TV is to create content for which people are willing to pay. This may entail completely new forms of content or conventional programs offered in new ways.

Advertising has been the major source of revenue for broadcast television and it may also eventually support Internet TV. But despite the promise of choice demographics, transaction tie-ins, and individual viewer targeting, advertisers still do not see a sufficient number of consumers watching video programs on computers to justify spending their dollars on Internet TV. Other potential revenue sources include promotional programs, or subsidies by established media institutions that are seeking to establish themselves in this new field. This last source might well be the economic foundation of Internet TV for some time.

Michael L. Katz of the Haas School of Business at Berkeley addresses business issues, beginning with industry structure. Katz seeks to determine how Internet TV might function in a world without distribution bottle-necks. He focuses on how Internet distribution will impact the television industry and points out four technological trends that will effect competition in each segment of the television value chain. He looks at vertical integration and bundling and analyzes their consequences for consumers and TV companies. He concludes that consumers will be the winners and local broadcasters will be the losers.

David Waterman of Indiana University identifies five economic characteristics of Internet technology and shows how they will lead to greater efficiencies. Economic models for content providers are developed based on his examination of the economic characteristics of Internet technology and how greater efficiencies are likely to shape them. Waterman uses historical evidence to predict that niche programs will flourish on the Internet but that high production value programming, particularly Hollywood movies, will continue to have a broader appeal. His implication is that, despite its potential reach, Internet TV may not have the cultural impact Hollywood enjoys.

Bertram Konert of the European Institute for the Media points out that new entrants and audience fragmentation are changing traditional media markets. In the long run, as bandwidth penetration increases, Konert sees Internet TV providing the new TV distribution system and recognizes the broadcasters' need to explore this distribution channel. Nonetheless, he argues for broadcasters to focus their efforts on developing multiple, self-supporting revenue streams and reminds them that return on investment is more important in the long term than the number of web impressions. Konert also warns against the continued subsidization of unprofitable engagements with revenues from profitable divisions.

## POLICY

This section of the book looks at policy issues, including copyright and regulation. It begins with a discussion of the regulatory climate for Internet TV in the United States and expands to a global perspective. The role of public service broadcasting is reviewed. One concern is that an information gap will emerge in society. Other issues involve content censorship, access regulation, cultural protection, copyrights, and compulsory licensing. The way that governments will regulate Internet TV is an open issue. Copyright laws may have to be adjusted.

Government regulators in the United States and elsewhere are looking at what role they should play in Internet TV. With too much regulation, new entrants may be excluded from creating and distributing new kinds of Internet TV content. But, left to themselves, problems may emerge, such as monopolization, consumer fraud, and programs potentially harmful to some viewers. There is also the question of which rules apply. Several traditional regimes for regulating telecommunications, broadcast, cable TV, and the print press exist. It is not likely that migrating one or another of these to Internet TV will work.

Robert Pepper of the Federal Communications Commission takes a look at the consumer content industries in the United States and on wholesale transmission. He sees a credible group of new entrants investing substantial dollars in infrastructure, and is encouraged that competition is well on its way. Pepper explains the past and present regulatory regimes. He points out the current dilemma concerning which regulatory model to use, if any, for Internet TV. He believes that complex intellectual property issues form the biggest barrier to Internet TV.

Christopher Marsden of the Economic and Social Research Council (ESRC) Globalization Center at the University of Warwick identifies and explores seven "pillars" of video over the Internet, namely, security, property rights, revenue, quality, access, standards, and competition. He contends that it is necessary to have a more legally certain international allocation of property rights to secure content from unauthorized use. Marsden also provides a review of current and developing technical standards. He concludes by examining competitive and oligopolistic scenarios for national and global market development.

Kenneth Carter of the Columbia Institute for Tele-Information points out the difficulties in implementing boundary-less Internet TV content in a television distribution environment divided into geographically defined broadcast regions. He reviews the history of the growth of this distribution network from its roots in analog radio to the present. Carter focuses on the intellectual property and distribution rights issues facing Internet delivery of video content. He points to potential consequences for content distributors should they fail to grasp the opportunities offered by the Internet.

Michael Einhorn of William Patterson University and CITI examines broadcast content copyright and licensing issues arising from new creation and distribution opportunities on the Internet. He believes that cyber-programmers will be able to time and space shift, edit, personalize, and repackage digital content at will to create new programs and venues. Einhorn acknowledges that the U.S. Copyright Office is correct in its determination that compulsory licensing for pricing retransmission is inappropriate. He then goes on to suggest, however, that such licensing might be appropriate in specific instances.

Fritz Pleitgen of the major German broadcaster, WDR, believes that it is the role of public service broadcasting to be the communications platform for all. He relates the efforts of German public broadcasting and its dual TV digitization and Internet initiatives that reflect their position on the future coexistence of the Internet and broadcasting.

Stephen Whittle of the British Broadcasting Corporation addresses regulatory issues and global inequality. He stresses the differences between illegal, unlawful, and harmful content and how to deal appropriately with each. He addresses the global issue of haves and have-nots—the "digital divide." Whittle speculates that an interest in the public, particularly through education and culture, might be broadband's means to success.

## CONTENT AND CULTURE

This section looks at what content is available, who is creating it, and how consumers view Internet TV content. Television viewing has been a passive experience in which viewers, often in groups, sit back and relax in their seats. Computers, in contrast, encourage a leaning forward posture, require more participation, and is usually a solitary activity.

John Carey of Greystone Communications examines the consumer demand for Internet TV. His research looks into the homes of ordinary Americans to show how technology is being used and how people's lives and homes have adapted to accommodate it. Carey notes that the use of computers in the home, coupled with the new opportunity to view Internet TV on these machines, has changed the way people organize their living spaces. His seeks to understand how people will choose to view television programs in the future based on changing consumer behavior today.

Jeffrey Hart of Indiana University asks whether Internet TV will be more of the same old TV. He explores the content companies and their business models, as well as the types of Internet TV content created. Hart divides Internet TV providers into several categories. He analyzes both new forms of intermediation made possible by Internet delivery and the strategies of leading streaming media players.

Gali Einav of Columbia University provides a look at the broad range of content companies that create and distribute interactive TV programs over the Internet. Her study gives an overview of the content side of this industry. She defines and examines the content provider landscape, which includes television broadcasters, Hollywood studios, independent producers, content syndicators and licensors, and user-generated content. Einav shows that Internet TV content models are based on both original and repurposed programming. Many of the early content companies did not survive. Constant adaptation to the environment has been essential to the rest. The traditional television content developers, including the TV networks, sustained losses and encountered technological limitations.

## GLOBAL IMPACTS

There are differences between how Internet TV is being introduced and received in the United States and in Europe. Many factors contribute to these differences, including technology and standards considerations, as well as Internet adoption and usage patterns that vary by country. This last section takes a look at some future global prospects for Internet TV content creation and distribution.

Eli Noam, Director of the Columbia Institute for Tele-Information, concludes the book with a discourse on the role the United States continues to play in television content creation and global distribution. To answer the question of the role of Hollywood in global content, Noam examines first the cost structures for several major different media: broadcast, theater, film, broadcast cable, TV, and Internet video. Among the electronic media, Internet TV is by far the most expensive method for distributing content. It is also fairly expensive to produce content for it. This leads to an analysis of what kinds of video content might be distributed over the Internet. These content types are premium VOD movies, specialized programs for dispersed niche markets, and interactive multimedia programs. The latter, in particular, also requires large production budgets, marketing skills, and advanced technology. Because the United States excels in supplying these components, Noam believes that it will continue its dominance in content production and distribution for traditional media with Internet TV. But, he points out, this U.S. dominance may trigger cultural and trade wars over Internet TV issues that will affect the move to an individualized Internet TV.

Thus, major obstacles must be overcome if Internet TV is to succeed. The road to its success is far more complex than the one from over-the-air TV to cable. Sufficient transmission capacity at the global, national, local, and home levels must be provided. Practices long established in the broadcast and cable industries must adapt. New business models and an adjustment in the value chain for TV content distribution are needed. Original Internet TV content productions must take advantage of the unique properties of computers and the Internet. Consumer behavior needs to evolve toward an acceptance of paid Internet content, and sufficient audiences must emerge. Regulators need to adjust their rules to a TV medium that is not based on scarce spectrum. The difficulties in achieving these major changes, in contrast to the hype surrounding it, have contributed to a slow start for Internet TV.

It may take awhile before both traditional and new kinds of broadcast quality television programming are regularly delivered and viewed over the Internet. This provides an opportunity to do a better job creating this new medium than was done with traditional TV broadcasting. It means there is a chance to get the economics, culture, technology, and policy right. Internet TV provides an exciting chance to return to the creativity of a new beginning.

# Infrastructure Implications of Internet TV

# 1

# Internet Television: Definition and Prospects

A. Michael Noll
University of Southern California

The Internet has been increasingly used during the past few years as a means to listen to radio shows in real time and to download audio. The downloaded audio could be music recordings (although copyright infringement has become an issue) and recordings of meetings and talks. The downloading of audio may extend in the near future to include the widespread downloading of video programming, perhaps even in real time, or what today is called video streaming.

Broadcast television is increasingly digital (Noll, 1988, 1999), so the transmission of video in digital form over the Internet on a broad basis may be appealing. Meanwhile, the display technology of the TV receiver and the personal computer are already similar, and TV sets are increasingly utilizing the processing power of computer technology. Thus, the technologies of the TV set and the personal computer are converging.

All paths appear to converge on Internet television (TV), as depicted in Fig. 1.1. Perhaps this convergence should be called "The Internet Meets Hollywood!" or "Video Meets the Internet." It is clear that the term "TV Over the Internet" is probably too narrow because it implies the use of a packet-switched data network to deliver conventional television video. Internet TV could be much broader and all inclusive than just this narrow definition. All these various forces focusing their energies on Internet TV might create an implosion of activity in the future, as shown in Fig. 1.2, or

1

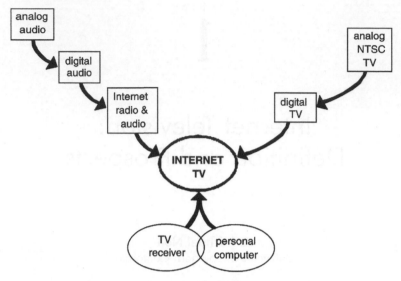

FIG. 1.1.   Evolutionary paths converge on Internet TV.

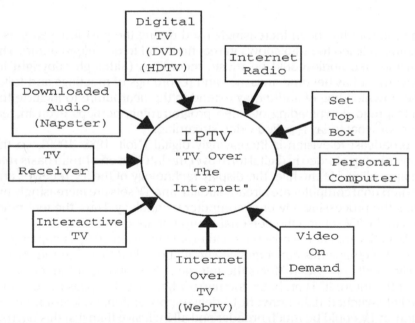

FIG. 1.2.   All the forces and developments in computers and television centralize on Internet TV. Whether all these forces focusing their energy on Internet TV will cause an implosion or not remains to be determined.

all the activity might create so much confusion and diffusion that some real opportunities could be lost along the way.

## INTERNET TV DEFINED

What is Internet TV? The definition depends on the definitions of television and Internet. Although the two terms might seem quite clear, they both are actually evolving with different meanings to different people.

In 1945, everyone knew that television was the broadcasting of moving images, along with sound, over the airwaves to homes for viewing on a television receiver. The content of these early days of television was news, movies, drama, sports, and variety shows. Later, television evolved to include the distribution of video content over coaxial cable, videotape, and videodisc. Although the distribution technology evolved, the content remained mostly unchanged.

Today, the media continually touts that "the Internet is everything." With consumers routinely downloading audio recordings and programs over the Internet, the next wave will be the routine, widespread downloading of video in real time over the Internet—a technology known as video streaming. Alternatively, video could be sent as a large file transfer and viewed at a later time. Either way, this could be the death knell of conventional television broadcasting, the video rental store, and such physical media as the digital videodisc (DVD). Consumers would then be able to pick and choose any video and television program for downloading and viewing at their convenience. The old dream of video-on-demand (VOD) will finally be realized.

But the network capacity required to transmit lengthy high quality video material to tens of millions of consumers will be costly and complicated to develop. The Internet is already used to download video shorts and cartoons. Some of this new video material is interactive, creating a new form of interactive television (ITV).

The definition of Internet TV obviously depends on the definition of the Internet. At the lowest infrastructure level, the Internet is a packet-switched data network consisting of transmission links and packet routers, linking computers around the globe. The Internet also means the protocols used for specifying the transmission and routing of packets of information. The Internet is the browser, such as Microsoft® Explorer® and Netscape® Communicator™, used to access and display the information obtained from Web sites on a personal computer in a friendly and easy-to-use fashion. The Internet is the general concept of the hypertext markup language (html) that is used to allow easy linkages between different sources of information along with friendly display. And lastly, the Internet is the look and feel of the information accessed and displayed in color and graphics from various Web sites. Clearly, the Internet is many things.

The various key features of the Internet are as follows: look and feel (color/graphics), browsers, hypertext markup language, Internet protocol for packet switching, and worldwide packet network for data transmission. What this all means is that the definitions of both the Internet and television are evolving. Hence, the definition of Internet TV, because it combines elements of both, is also evolving and is not yet clear. Therefore, the technology and future of Internet TV are somewhat speculative because of this unclear definition. If the definition of the "it" is unclear, then most consumers will not buy "it," and the "it" will have problems in the marketplace or evolve into an "it" that is less hazy and more clearly defined.

## INTERNET TV CAN BE MANY THINGS

Internet TV is many things, or even a combination of things. In its most obvious implementation, Internet TV is conventional television obtained over the Internet. Rather than watching television programs broadcast over the air or over cable, television programs are accessed over the Internet and then watched in real time, using a technology known as video streaming. Not only conventional television, but also cartoons and video shorts, are sent over the Internet with video streaming. All this video is watched on the personal computer. Computer technology will be incorporated within future television sets to facilitate television access over the Internet. The television set thus converges with the personal computer.

Internet TV is the adoption of an Internet-like interface in accessing and watching television—a new form of video navigation over the Internet. Internet TV is a more interactive approach to controlling the television experience with the ability to obtain all sorts of ancillary information while watching television, as promoted by Wink Communications.

Internet TV is the use of the home TV set to view Internet sites, as offered by WebTV Networks, perhaps in conjunction with conventional television viewing. These kinds of applications of Internet TV create an interactive television experience called Internet-enhanced TV. Such Internet-enhanced TV could then evolve into Internet-delivered TV on a wide basis.

Internet TV is the use of the Internet protocol to store and transmit video, both at the TV studio and also to various locations. Rather than storing and transmitting digital video as a continuous stream of bits, the digital video is packetized into the packets specified by the Internet protocol.

## EVOLUTIONARY CONVERGENCE

The personal computer is the device in the home used to access the Internet through a modem connected either to a telephone line or to a coaxial cable. The personal computer is situated at the home work center. Television is watched on the home television receiver, physically placed at the home entertainment center. Will the personal computer evolve into an entertainment center? Will the home television receiver become the means to access the Internet, as promoted by Web TV? Or will television receivers of the future increasingly adapt digital technology for entertainment purposes, leaving the personal computer as a separate work appliance in the home?

## LESSONS FROM WEB RADIO

Radio is being changed because of the Internet. Will similar changes occur to television? Consumers in rural areas now "listen" to the radio over the Internet. A colleague in rural New Hampshire accesses WQXR in New York City over the Internet to listen to classical music. Will radio stations return their broadcast licenses as they migrate to the Internet?

Indeed, there is growing evidence of a much broader market for radio stations beyond their local market. But there are more technologically efficient ways to extend this reach than the use of the packet-switched Internet. One way would be geostationary satellite transmission of all radio stations. Just two satellite transponders could handle 10,000 radio stations in a compressed digital format. But would small inexpensive satellite radio receivers be developed?

Internet TV could then simply evolve into a form of "world TV" in which all television programming from the entire planet would be transmitted by satellite to everyone on the earth. The capacity to do this would require only a few geostationary satellites.

## A LITANY OF FAILURES, YET CAUSE FOR HOPE

An issue of *The Economist* characterizes "the digital revolution in entertainment ... [as] somewhere between a disappointment and a disaster" (Duncan, 2000). Indeed, much of the rhetoric of Internet television is very similar to the words of the recent past about such technologies as video dialtone (VDT), video-on-demand (VOD), interactive TV (ITV), and fiber to the home (FTTH). Unfortunately, these technologies are a litany of failures, along with such others as Web TV and high-definition digital TV.

The stereotypical image of the television viewer as a couch potato is based much on reality. Most TV viewers do not want to interact with their TV sets, other than to click the remote to change channels (Hansell, 2000). However, the British and many Europeans do interact with their TV sets via teletext—as yet, the only successful form of interactive television (Noll, 1985). Teletext transmits a few hundred frames of text and graphics during the vertical blanking interval of a television signal. Because of confusion over standards, teletext never developed in the United States. The success of the Internet, which offers access to far more information than teletext, means that teletext has a dim future.

Interactive two-way television was pioneered decades ago by Warner Cable in Columbus, Ohio. The Warner QUBE system was initiated in 1977, but was later terminated in the late 1980s. Although all sorts of wild promises were made for its application, a major use of QUBE was to obtain instant audience reaction to new programs.

Web TV offers Internet access over the home television receiver using a keyboard that attaches to the telephone line and to the TV set for display. The service has met with very little consumer acceptance ("Whatever Happened to WebTV?," 2000). Yet, the predictions of a huge market for interactive television continue to be made, with Forrester Research predicting over 20% of TV revenue from smart set-top boxes by 2005 ("Microsoft's Blank Screen," 2000).

GTE offered all sorts of interactive, two-way, broadband video in Cerritos, California, from the late 1980s to the late 1990s. The consumer interest simply was not present for the video-on-demand (VOD) service. In 1992, Bell Atlantic promised to deploy optical fiber in the local loop for video dialtone (VDT) and predicted to have 1.2 million homes connected by end of 1995. Nearly 6,000 homes in Dover Township, New Jersey, were connected to an optical fiber system in 1996, only to be disconnected a few years later. The service was apparently far too costly to continue to offer. Bell Atlantic has never explained the reasons for withdrawing the service.

High-definition television (HDTV) doubles the number of scan lines of conventional television. HDTV is being used in the TV studio to capture video in a high quality format. HDTV is also being broadcast in a digital format in an additional channel in the UHF spectrum given to conventional VHF broadcasters. Consumers, thus far, have shown little interest in HDTV. A key issue will occur over what to do with all the UHF spectrum given to the broadcasters for a service that is little watched.

The telephone network is a switched audio network. It could have been used as a means to call radio stations to listen to programs on a switched basis. But the switched public telephone network is not used that way, perhaps because of the cost. Listening to radio programs over the telephone network at 5 cents per minute equates to $3 per hour, which is quite prohibitive. The advantage of the Internet for doing the same thing is that the Internet is "free" to consumers and does not have usage sensitive pricing.

The concept of broadband to the home is not new and was first presented in the 1970s. The "broadbandwagon" keeps rolling along, reinventing itself every few years (Noll, 1989). A decade ago, it was fiber to the home (FTTH) and fiber to the curb (FTTC). Today, it is Internet TV, but there is a history of failures of this "highway of dreams" of an information superhighway (Noll, 1997a). Those who are able to maintain hope against the overwhelming tide of failures will say the time is now, but it might be still just too soon.

## UNCERTAINTIES ABOUND

There are technological challenges and hurdles that would need to be overcome for Internet television to become a reality. However, the real uncertainties are not technological in nature but involve consumer needs and behavior, industry structure, and finance. In addition, there are issues over copyright and ownership protection that would need to be resolved.

The Internet is mostly "free" except for a fixed monthly access charge paid to the Internet access provider. But if there were tremendous use of the Internet to send huge amount of video data, usage sensitive pricing might be needed (Noll, 1997b). Even a usage charge as low as $0.1$ $\mu$¢/bit $(0.1 \times 10^{-6}$ cents per bit) adds up quickly when video data is sent continuously at 4 Mbps. A minute of such Internet high quality video would incur usage charges of 24 cents, which is prohibitively expensive. However, there are many other uncertainties about Internet television.

Are consumers demanding Internet television? Do consumers want to interact with their television sets? Is it possible to overcome the history of the past failures and false promises of video dialtone, video-on-demand, interactive TV, and fiber to the home?

Is Internet TV a revolution or will it become yet another evolution in the distribution of video programming in digital form? As bandwidth continues to become available, will consumers demand improved quality so that compression is no longer wanted or needed? Is Internet TV a way to be entertained or a way to obtain information? How will conventional broadcasters respond? Is Internet TV just an evolutionary enhancement of digital CATV? Will new unexpected applications evolve and change the Internet TV landscape? What will Internet TV cost, who will finance it, and who will afford it? The answers to these key questions are currently unclear. Thus far, Internet TV has been driven mostly by technology push. Will consumer pull develop, thereby leading to success? Only time will tell.

## ACKNOWLEDGMENT

Many of the definitions of Internet TV presented in this chapter are a result of discussions with Darcy Gerbarg, and her contributions to this chapter are gratefully acknowledged.

## REFERENCES

Duncan, E. (2000, October 7). E-entertainment thrills and spills. *The Economist,* p.6.

Hansell, S. (2000, September 20). Clicking outside the box. *The New York Times*, p. H1.

Microsoft's blank screen. (2000, September 16). *The Economist*, p. 74–75.

Noll, M. A. (1985, ). Videotex: Anatomy of a failure. *Information & Management, 9*(2), 99–109.

Noll, M. A. (1989, September). The broadbandwagon. *Telecommunications Policy, 13*(3), 197–201.

Noll, M. A. (1991). High definition television (HDTV). In A. A. Berger (Ed.), *Media USA* (2nd ed., pp. 431–438). New York: Longman.

Noll, M. A. (1997a). *Highway of dreams: A critical appraisal of the communication superhighway*. Mahwah, NJ: Lawrence Erlbaum Associates.

Noll, M. A., (1997b). Internet pricing vs. reality. *Communications of the ACM, 40*(8), 118–121.

Noll, M. A. (1998). The digital mystique: A review of digital technology and its application to television. *Prometheus, 16*(2), 145–153.

Noll, M. A. (1999). The evolution of television technology. In D. Gerbarg (Ed.), *The economics, technology and content of digital TV* (pp. 3–17). Boston: Kluwer.

Whatever happened to WebTV? (2000, September 16). *The Economist*, p. 75.

# 2

# Implications for the Long Distance Network

Andrew Odlyzko

*University of Minnesota*

Traditional concerns about the impact of television on the long distance links in the Internet are unjustified. The convergence of TV and the Internet is likely to be slower and take a path different from the one normally envisioned. There are various definitions of Internet television. (See Egan, 1996; Noll, chaps. 1 and 3 in this vol.; Owen, 1999.) But, the precise one does not matter much for the purposes of this chapter.

Data networks developed rapidly largely because they could use the huge existing infrastructure of the telephone network. Without all the investments made to provide voice services, long distance data transmission would have grown much more slowly. As it is, growth has been fast, although not as fast as is commonly believed. The bandwidth of data networks in the United States already exceeds that of voice networks (see the next section for more details). Sometime in 2001 or 2002, the volume of data transmitted on data networks exceeded that of voice. (See Coffman and Odlyzko, 2001b, for the historical growth rates of different types of data and nondata traffic, and of predictions of when data would exceed voice.)

Despite all the publicity it attracted in the late 1990s, packetized voice is still a tiny fraction of Internet backbone traffic. This is not likely to change, even as a greater fraction of voice is sent over the Internet. The reason is the far higher growth rate of data traffic than of voice calling, about 100% versus under 10% per year. Even today, to move all current voice traffic to the Internet would require much less than a doubling of the Internet's capacity.

Although there are still about as many bytes of voice traffic as of Internet traffic, packetization of voice naturally lends itself to compression. Hence, if voice traffic were to move to the Internet, then the volume of packet traffic that would result would be far smaller than current data traffic.

The general conclusion is that the volume of voice calls is not going to overwhelm the Internet. (There are quality issues as well that matter, but this chapter does not deal with those here.) Historically, however, data networks have developed in the shadow of the telephone network. Not only have data networks relied on the infrastructure of voice telephony, but their development was strongly influenced by the prospect that eventually they would carry voice. Because the telephone network was so much larger than the data networks, the quality requirements for voice transmission played a major role in the planning of data transmission technologies. Right now, it is increasingly realized that voice will not be a large part of the traffic in the future, simply because there is too much data. On the other hand, video is now playing a similar role to the one voice used to play. The volume of TV transmissions is so large that the requirements of real-time streaming video dominate planning for the future of the Internet. However, that is also likely to turn out to be a mistake. By the time TV moves to the Internet, data traffic will likely be so large that streaming video will not dominate it. Moreover, the video traffic on the Internet is likely to be primarily in the form of file transfers, not streaming real-time transmission.

The aforementioned contrarian predictions are based on a study of rates of change in different fields. Storage, processing, display, and transmission technologies are advancing at rather regular and predictable rates. This is considered in later sections. ("Moore's Law" for semiconductors is only the most famous of the various "laws" that govern progress.) In addition, rates at which new technologies are adopted by society, although not as regular, are almost universally much slower than is commonly supposed. ("Internet time" is a myth.) This is discussed at greater length later. As a result, there is some confidence in expecting that by the time TV moves to the Internet in a noticeable way, the latter will have huge capacity, at least on the long distance links.

The prediction that the predominant mode of video transmission on the Internet is likely to be through file transfers is justified briefly later. The concluding section is devoted to what appears to be the most likely impact of the Internet on TV, namely, in providing greater flexibility that will encourage exploration of technologies like high-definition television (HDTV).

## NETWORK SIZES AND GROWTH RATES

Coffman and Odlyzko (1998) pointed out that already by the end of 1997, the bandwidth of long distance data networks in the United States was comparable to that of the voice network, with the public Internet a small fraction of the total. Today, the Internet is by far the largest in terms of bandwidth. However, because bandwidth is hard to measure and changes irregularly, due to the lumpy nature of network capacity as well as the financial climate, it is hard to estimate it precisely. Table 2.1 presents the estimate from Coffman and Odlyzko (2002b) of the traffic (in terabytes, units of $10^{\wedge}12$ bytes, per month). The key point, discussed in great detail in Coffman and Odlyzko (1998, 2002b) is that Internet traffic is growing at about 100% per year. That is the growth rate the Internet experienced during the early 1990s. There was then a brief period of 2 years (1995, 1996) when growth was at the "doubling every three or four months" rate that is usually mentioned. Starting in 1997, however, growth again slowed down to doubling each year. At this rate, by some time in 2001 or 2002, there was more data than voice traffic in the United States, as predicted in Coffman and Odlyzko (1998).

Further, technology advances in transmission and switching appear to offer the prospects of traffic growing at about 100% a year through 2010 without astronomical increases in spending. Even if growth occurs by a cumulative factor of 100 over the first decade of the 21st century (as opposed to a factor of 1,024 that results from a doubling each year), there will be around 3 million TB per month of traffic by the end of 2010, or around 10 GB per person per month. Now a 90-minute movie, digitized for high resolution at 10 Mb/s, comes to about 7 GB, so it would be possible to transmit only about two movies per person (counting all men, women, and children) per month in that format. However, if the resolution is lowered to 2 Mb/s, and it is assumed that traffic continues doubling each year, by 2010 it would be possible to send 100 movies per person per month. Thus, the general conclusion is that by 2010, or soon thereafter, the long distance Internet backbone could transmit all the entertainment TV signals that are likely to be demanded.

Another way to consider the problem of the transportation task imposed by TV is by considering capacities of fibers. If each of the approxi-

TABLE 2.1
*Traffic on U.S. long distance networks (year-end 2000)*

| Network | Traffic (TB/month) |
|---|---|
| U.S. voice | 53,000 |
| Internet | 20,000–35,000 |
| Other public data networks | 3,000 |

mately 300 million inhabitants of the United States is given a 10 Mb/s traffic stream, the total demand would be for 3,000 Tb/s of transmission capacity. The dense wavelength division multiplexing (DWDM) technologies that are widely deployed typically reach about 0.8 Tb/s per fiber strand, but there are good prospects of reaching 10 Tb/s in a few years, and there are even hopes of achieving 100 Tb/s.[1] If it is assumed conservatively that 10 Tb/s capacity per fiber will be widely deployed by 2010, then it would require just 300 strands to provide the 3,000 Tb/s of capacity that the 10 Mb/s traffic stream per person involves. (Actually, double that would be needed for two directions of traffic, plus other small multiplier factors to provide for redundancy, etc., but those are not huge factors.) Today, there are several hundred strands of fiber running from coast to coast, and many empty conduits that could be filled with additional fiber. Thus, as far as fiber itself is concerned, there will be plenty of capacity.

Most of the fiber that is in place in the long distance networks is not utilized ("lit" in industry language), and even when it is in use, it is often used at a small fraction of its capacity. The reason is that there is not enough demand to create more usable capacity, certainly not even at the prices of 2001 (which are much lower than they were just a couple of years ago). (The present fiber glut resulted from an assumption that there was an insatiable demand for bandwidth. It ignored three key factors: lack of "last mile" connectivity; the cost to provide usable bandwidth, as opposed to raw fiber; and, perhaps most important, traffic demand is growing at only about 100% per year, even in the absence of bandwidth constraints, gated more by rate of adoption of new applications than anything else.)

The general conclusion is that there already is enough fiber to allow for transmission of individual TV signals over the long distance Internet backbones, and sometime around 2010, transmission and switching technologies are likely to allow for this to be done economically. The question is, will we want to do that? The volume of unique TV content is simply not all that large, as is shown in Lesk, (1997) and Lyman and Varian (n.d.). Given the trends in storage capacity mentioned later, it is feasible to store copies of all the nonreal-time material (which is the overwhelming bulk of what TV transmits) on multiple local servers, and avoid burdening the backbones with it.

## MOORE'S LAWS (TECHNOLOGY TRENDS)

In the previous section, there was an implicit assumption, namely, that the highest resolution video signals that would be typical by 2010 would be no more than 10 Mb/s. Today, on digital cable TV systems, typical transmission

---

[1]DWDM is a technology for increasing the bandwidth of an individual fiber strand by sending many signals simultaneously, each on a different wavelength.

rates are around 2 Mb/s, and HDTV signals tend to be compressed to somewhat below 10 Mb/s. Increases in resolution of video signals can certainly be expected. (Movies are filmed at over 1 Gb/s, and stored as such.) However, these increases are likely to be modest. (Note that TV resolution has not changed in over 50 years, and HDTV and other forms of enhanced display technologies have been making slight progress, a point to be considered further later.)

In general, technological prognostications have a miserable track record. The one area where they have been outstandingly successful, however, has been in forecasting continuation of various types of laws similar to the "Moore's Law" of semiconductors, which says that the number of transistors on a chip doubles every 18 months. (See Schaller, 1997, for the history and fuller description. The basic law is often reported as stating that processor power doubles every 18 months, which is not quite right, but reasonably close.) The key point, discussed at greater length in Coffman and Odlyzko (2002b), is that the different Moore's laws for different areas operate at different speeds. Display resolution is improving slowly (and battery capacity even more slowly), and transmission and magnetic storage capacity are growing even faster than processor power. Table 2.2 shows the growth in the volume of hard disc storage that is shipped each year. It is about doubling annually, comparable to the rate at which transmission capacity is growing.

The rapid growth of storage capacity is significant, because it makes nonstreaming modes of operation much more attractive. Back in the 1980s and 1990s, disc storage available on PCs in households was so small that streaming real-time delivery of video was the only feasible alternative. Today, local storage is becoming viable even for high resolution movies. (Note the estimate of 7 GB for a single HDTV movie, versus a capacity of 80 GB that often comes with high-end PCs in mid-2001, and the likelihood that this will reach 1 TB around 2005). As time goes on, and the disc capacity grows rapidly while digital movie sizes grow slowly, the attractions of local storage will only increase.

## RATES OF CHANGE, TECHNOLOGICAL AND SOCIOLOGICAL

People hear constantly how they live on "Internet time," and how the Internet changes everything. Yet "Internet time" is a myth. The pace at which new products and services are adopted is not notably faster than it used to be in the past. This contrarian view is considered in greater detail in Odlyzko (1997). Because it is so contrarian, some space here is devoted to justifying it (and presenting more examples).

There are frequently cited graphs showing faster diffusion of new technologies today than a century ago, say, such as those in Cox and Alm (1996). However, those comparisons have to be treated with caution.

TABLE 2.2
*Worldwide Hard Disc Drive Market*

| Year | Revenues (Billions) | Storage Capacity (Terabytes) |
|------|---------------------|------------------------------|
| 1995 | $21.593 | 76,243 |
| 1996 | 24.655 | 147,200 |
| 1997 | 27.339 | 334,791 |
| 1998 | 26.969 | 695,140 |
| 1999 | 29.143 | 1,463,109 |
| 2000 | 32.519 | 3,222,153 |
| 2001 | 36.219 | 7,239,972 |
| 2002 | 40.683 | 15,424,824 |
| 2003 | | 30,239,756 |
| 2004 | | 56,558,700 |

*Note:* Based on September 1998 and August 2000 IDC reports. Table from Coffman and Odlyzko (2001b).

Yes, the telephone, the automobile, and electricity did spread slowly, but then each had to build its own extensive infrastructure, and each one was very expensive in its first few decades. The Internet could take advantage of the existing telephone network to grow, and yet even the Internet did not really grow on "Internet time," because its origins go back to the Arpanet, which was put into operation in 1969. For successful new consumer products or services that do not require large investments, a decade appears to be about the length of time it takes for wide penetration. This was noted a long time ago.

A modern maxim says:

> People tend to overestimate what can be done in one year and to underestimate what can be done in five or ten years. (Licklider, 1965, p. 17)

Arthur C. Clarke, the science fiction writer, is said to have similarly claimed that people tend to overestimate the short-term impact of new technologies and to underestimate the long-term impact.

Color TV took about a decade to reach 75% of the households in the United States. It is not much different today. Odlyzko (1997) presented statistics on sales of recorded music in the United States by format. Music CDs are much better than vinyl LPs (at least for 99% of the population, although there is a small but influential segment that insists on the superiority of the older medium), yet it took them around a decade to attain dominance. Cell phones are all the rage, but they have been around since the mid-1980s, and yet by the end of 2000 they were used by just about 40% of the U.S. population.

The standard example of how things do move on "Internet time" is the browser. It did attain dominance in providing online access in under 2 years. But that is just about the only such example of rapid change! Even on the Internet, technologies such as IPv6 and HTTP1.1 have been talked about as the "next big thing" for about half a dozen years, and are not yet dominant. Amazon.com did revolutionize retailing. However, it took quite awhile, because it was established in November 1994, and 6 years later it had not yet taken even 10% of the U.S. book market. (Whether Amazon.com is viable in the long run or just an outstanding example of the "irrational exuberance" of the financial markets is another story.)

Much of the dot-com bubble appears to have been due to the expectations that the world was changing on Internet time. For example, in the middle of 2001, just before Webvan closed down, its new CEO was quoted as saying, "We made the assumption that capital was endless, and demand was endless." The idea of deliveries to the home may yet find a market and lead to financial success. However, Webvan was acting under the assumption that they had to build a giant distribution network in a year or two, or else somebody else would. Instead, when demand was slow to materialize, they went bankrupt.

The entertainment area is full of examples of slow changes. Galbi (2001) provided interesting statistics on a variety of subjects. Some of the most relevant have to do with the slow rate at which people reallocate their time. For example, reading went from 4 hours per week to 3 hours, but this change took from 1965 to 1995, a 30-year period, to occur. More examples of slow consumer adoption rates are appearing all the time. For example, personal video recorders, such as TiVo and ReplayTV, have so far failed to take off, even though their users praise them highly (Hamilton, 2001).

The general conclusion is not to expect much change in consumer behavior as far as entertainment is concerned, at least not in less than 10 years. In particular, TV is likely to retain its format, and be delivered through TV sets, not PCs. In the meantime, the backbones of the Internet will be growing, to the stage where they will be capable of delivering all the TV content as separate streams for individual users even from a single central location. That mode of delivery is irrationally inefficient, so it is unlikely to be employed, and TV signals consequently will not fill much of the Internet pipelines.

## STREAMING MEDIA VERSUS STORE-AND-REPLAY

Where will the increases come from if Internet traffic continues doubling each year? There are some speculations in Coffman and Odlyzko (2002a). Video is likely to play an increasing role, taking over as a major driver of traffic growth from music (which got a large boost from Napster). However, this video is likely to be in the form of file transfers, not streaming real-time traffic. There are more detailed arguments in Coffman and Odlyzko (2002a), but the basic argument is that video will follow the example of Napster (or MP3, to be more precise), which is delivered primarily as files for local storage and replay, and not in streaming form. It has been known for a long time that this local storage and replay model is a possibility (cf. Owen, 1999). It has several advantages. It can be deployed easily (no need to wait for the whole Internet to be upgraded to provide high quality transmission). It also allows for faster than real-time transmission when networks acquire sufficient bandwidth. (This will allow for sampling and for easy transfer to portable storage units.)

The prediction that streaming multimedia traffic will not dominate the Internet has been made before (Odlyzko, 2000; St. Arnaud, 1997). It fits in well with the increasing abundance of local storage.

## CONCLUSIONS

The general conclusion is that the long distance Internet backbones are not going to be affected much by TV. Local "last mile" bottlenecks in data networks, as well as the slow adoption rates of new technologies by consumers, will ensure that by the time true convergence takes place between the Internet and entertainment TV, something on the order of a decade will have gone by. By that point, the backbones will have more than enough capacity to handle TV transmission. Even though it may be wasteful, it may then very well be less expensive to handle everything over the Internet, to avoid having several separate networks.

The Internet may very well have a larger impact on TV than TV will have on the Internet. The main advantage of the Internet has always been its flexibility, not its low cost. (See the discussions in Coffman and Odlyzko, 2002b; Odlyzko, 2000.) The broadcast model, in which people have to adjust their schedules to fit those set by network executives, was an unnatural one, forced by the limitations of the available technology. The popularity of videotape rentals showed that people preferred flexibility. Similarly, when cable TV operators chose to offer more channels as opposed to higher resolution channels, they were presumably responding to what they saw as their customers' desires for variety.

The Internet will offer even more flexibility, but its impact is unlikely to be very rapid. Its main effect may be on high resolution video. HDTV has made practically no inroads because of the usual chicken-and-egg syndrome. Sets are expensive because there is no mass market, people do not buy sets because they are expensive and there is nothing novel to watch, stations do not carry HDTV programming because there is no audience, and so on. Internet allows for marketing to small groups. Studios already are making high resolution digital version of movies, and over the Internet will be able to reach the initially small groups of fans willing to pay extra for them. (This too will take time, not least because of fears of piracy.) Experiments with novel modes of presentation will also get a boost.

## REFERENCES

Coffman, K. G., & Odlyzko, A. M. (1998). The size and growth rate of the Internet, *First Monday*, October, 1998, http://firstmonday.org/ Also available at http://www.dtc.umn.edu/~odlyzko

Coffman, K. G., & Odlyzko, A. M. (2002a). The growth of the Internet. In I. P. Kaminow & T. Li, (Eds.), *Optical fiber telecommunications IV B: Systems and impairments* (pp. 17–56). Academic Press. Available at http://www.dtc.umn.edu/~odlyzko

Coffman, K. G., & Odlyzko, A. M. (2002b). Internet growth: Is there a "Moore's Law" for data traffic? In J. Abello, P. M. Pardalos, & M. G. C. Resende (Eds.), *Handbook of massive data sets*, (pp. 47–93). Kluwer. Available at http://www.dtc.umn.edu/~odlyzko

Cox, W. M., & Alm, R. (1996). *The economy at light speed: Technology and growth in the Information Age and beyond,* Annual Report, Federal Reserve Bank of Dallas, Exhibit D (p. 14). Available at http://www.dallasfed.org/ htm/pubs/annual.html

Egan, B. L. (1996). *Information superhighway revisited: The economics of multimedia,* Artech House. Updates available at http://began.com/broad3.htm

Galbi, D. (2001, July). Some economics of personal activity and implications for the digital economy, *First Monday, 6*(7), http://firstmonday.org/issues/issue6_7/galbi/

Hamilton, D. P. (2001, February 7). TiVo, ReplayTV fail to take off despite big fans, *Wall Street Journal.*

Lesk, M. (1997). *How much information is there in the world?*, unpublished paper. Available at http://www.lesk.com/mlesk/diglib.html

Licklider, J. C. R. (1965). *Libraries of the future,* MIT Press.

Lyman, P., & Varian, H. R. (n.d.). How much information? Report available at http://www.sims.berkeley.edu/how-much-info/

Odlyzko, A. M. (1997). The slow evolution of electronic publishing. In A. J. Meadows & F. Rowland, (Eds.), *Electronic publishing—new models and opportunities* (pp. 4–18). ICCC Press. Available at http://www.dtc.umn.edu/~odlyzko

Odlyzko, A. M. (2000). The Internet and other networks: Utilization rates and their implications, *Information Economics & Policy, 12*, pp. 341–365. (Presented at the 1998 Telecommunications Policy Research Conference.) Also available at http://www.dtc.umn.edu/~odlyzko

Owen, B. M. (1999, June). *The Internet challenge to television,* Harvard University Press.

Schaller, R. R. (1997, June). Moore's law: Past, present, and future, *IEEE Spectrum, 34*(6), 52–59. Available through Spectrum online search at http://www.spectrum.ieee.org

St. Arnaud, B. (1997, November). The future of the Internet is NOT multimedia, *Network World.* Available at http://www.canarie.ca/~bstarn/publications.html

# 3

# Television Over the Internet: Technological Challenges

A. Michael Noll
*University of Southern California*

Technology is a key factor in shaping the future, but there are many technological uncertainties and challenges that will need to be resolved for television over the Internet to become a reality. This chapter discusses these technological challenges.

## DELIVERY INFRASTRUCTURE

Television and video are delivered by a variety of means to consumers, as is depicted in Fig. 3.1. Conventional broadcast television is transmitted over the air by radio waves in the VHF and UHF frequency bands. Broadcast television originates at the television studio, and increasingly broadcast television signals at the studio are in a digital format, although conventional VHF/UHF transmission remains mostly analogue.

An additional channel in the UHF band was given to the VHF broadcasters to be used for the broadcasting of digital TV, which could be high-definition TV (HDTV) or the multicasting of a number of digital TV programs at conventional resolution. Consumer response to such digital TV has been very low thus far in the United States.

Television programs are transmitted to earth from communication satellites located in geostationary orbits above the Earth's equator by virtue of a technology called direct broadcast satellite (DBS) television. DirecTV is the dominant DBS provider in the United States. Over 200 television pro-

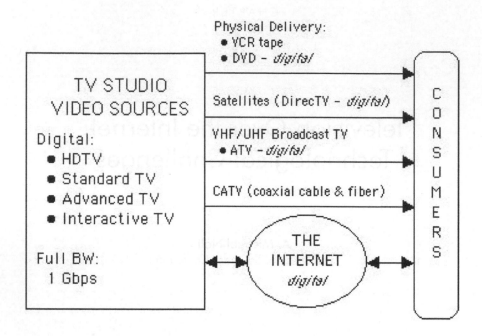

FIG. 3.1.   Television signals are delivered to consumers over a wide variety of media and means.

grams are converted to a digital format, compressed to reduce their bit rate, and broadcast to earth over the DirecTV DBS service. Over 15% of households in the United States obtain their television from DBS.

Another 64% of households in the United States obtain their television over the coaxial cable of the local cable television (CATV) service provider, with AT&T currently being the largest CATV provider in the United States. The signals sent over CATV are mostly analogue television at the conventional bandwidth of 6 MHz per TV channel. Given the success of DBS with its digital transmission of television, CATV will increasingly likewise migrate toward digital, thereby being able to offer more channels of programming. It would be nice if the quality of programming content were to match the increase in the number of channels, but it seems that more channels equates to lower program quality because each channel is able to capture a much smaller audience.

DBS, CATV, and VHF/UHF television signals are all broadcast electronically, either as radio waves or as electrical waves over coaxial cable. Recorded video is delivered physically on magnetic tape or on laser discs. Prerecorded video is used for mostly movies and for other programming.

Such video has been physically delivered over tape—the videocassette recorder (VCR). Increasingly, video is being delivered physically in digital form in a compact digital videodisc (DVD).

## COMPRESSION

At the TV studio, television cameras are still mostly analogue, but digital technology is becoming increasingly pervasive. Within the studio, digital television signals operate at full digital rates. HDTV digital requires about 1 Gbps ($10^9$ bps), and digital video recorders are available to record such fast rates. What is not available are media capable of transmitting such fast rates to consumers in their homes. Compression is the interim solution until more bandwidth becomes available.

Television signals consume vast amounts of bandwidth. An analogue broadcast television channel in the United States requires 6 MHz of spectrum space. The analogue video signal itself requires about 4 MHz of analogue bandwidth. Converting this analogue video signal to a digital format for home quality can consume a bit rate of about 84 Mbps. High-definition professional studio quality requires about 1 Gbps. Such tremendous bit rates are very costly to transmit over conventional broadcast and switched systems. The solution is compression.

Video images change slowly from frame to frame. Rather then transmit the entire information in each frame, compression techniques encode and transmit only the changes in information from one frame to the next. This is interframe compression. Within each frame, the information from one scan line to the next is very similar. Compression techniques encode the changes from one scan line to the next and also look for similar blocks of information. This is intraframe compression. Compression algorithms have been developed by the Moving Pictures Expert Group (MPEG) of the International Standards Organization (ISO).

A conventional television signal has an analogue bandwidth of about 4 MHz. When converted to a digital format, this television signal requires a bit rate in the order of roughly 100 Mbps. When compressed, a much lower bit rate in the order of about 4 Mbps suffices. Compression is, however, a compromise with quality, and certain artifacts can appear in the reconstructed television image. Most of these artifacts would not be bothersome to most viewers, however. Compression to lower bit rates as low as only 1 Mbps can be done if the degradation in quality is acceptable.

## TECHNOLOGICAL SPECIFICATIONS

There are a number of technological factors that characterize signals. Various factors are shown in Table 3.1 for voice, data, and video signals. The

TABLE 3.1
*Technological Factors*

|  | Voice | Data | Video |
|---|---|---|---|
| *Bandwidth:* | | | |
| • analogue | 4 kHz | | 6MHz |
| • digital | 64kbps | 56 kbps to 1 Mbps | 84 Mbps |
| • compressed: | | | |
| good quality | 8–12 kbps | | 4 Mbps |
| lower quality | 1.2 kbps | | 1 Mbps |
| Directionality | Two-way | Two-way | One-way |
| | Full duplex | Half duplex | Simplex |
| Holding time | Minutes | Minutes | Hours |
| Symmetry | Symmetric | Asymmetric | Asymmetric |
| Network | Switched | Switched | Broadcast |
| Timing | Distributed | Distributed | Prime time |
| Purpose | Communication | Information | Entertainment |

factors for these three signals are quite different, which might imply different kinds of networks to deliver each.

Perhaps the most important factor in characterizing a signal is the bandwidth—either analogue Hertz or digital bits per second—required to transmit the signal. In analogue form, telephone quality voice requires 4 kHz; an analogue television channel consumes 6 MHz. Analogue signals can be compressed to reduce their bandwidth. Analogue telephone speech could be reduced by a factor of 10 using vocoder technology.

Digital signals are characterized by their bit rates, but they also occupy analogue bandwidth. All signals, analogue or digital, occupy bandwidth. A telephone signal converted to digital format requires 64 kbps; a television signal requires 84 Mbps. The digital television signal can be compressed to as low as 1–4 Mbps, but a digital speech signal can be compressed to as low as 1.2 kbps. Whether analogue or digital, television requires considerably more bandwidth and bit rates than telephone speech. Data signals fall in between speech and video in terms of bit rate.

Most voice telephone calls are quite short, in the order of a few minutes, although some people visit for hours by phone. Traditional television programs and movies are much longer in duration, usually a half hour or more. Although most data sessions are short, downloading a large file can take a long time. Voice requires a symmetric switched connection with equal transfers of information in each direction. Video is broadcast in one direction. Voice requires a full duplex connection to enable both parties to speak simultaneously. Data directionality can switch in a half duplex manner. Video is usually one-way all the time in a simplex manner, except for interactive, two-way television. Data requires a switched network, but most of the data traffic is asymmetric in one direction.

The audience timing for voice and data communication is distributed during the day, although there are some peak hours of use. Video is usually watched by nearly everyone at the identical prime-time hours. The purpose of voice is communication, the purpose of data is information, and the purpose of video is mostly entertainment. Thus, voice, data, and video are quite different in terms of their technological characterization. This might imply different network architectures and delivery technologies.

## NETWORK ARCHITECTURE

Networks can be classified according to their locale. Backbone networks (called long distance networks for voice telephony) cover great distances, such as across continents or under oceans. Local access networks are the means in which access to backbones is obtained. Lastly, there are intrapremises networks that carry signals from one computer to another within a building or office. The technological challenges for Internet television are different for these different networks.

Networks consist of three major areas: transmission, switching, and control. Transmission deals with the various media over which signals travel from the source to the destination. Transmission also deals with how many signals can be combined to share a medium—what is called *multiplexing*. Switching and control treat the various methods for assuring a signal gets to a specific destination.

There are many transmission media: copper wire, optical fiber, coaxial cable, and radio waves. Signals can be multiplexed by assigning each its own unique band of frequencies, or what is known as *frequency-division multiplexing*. Digital signals can be multiplexed by assigning each its own unique place in the time sequence, or what is known as *time-division multiplexing*.

In the analogue world, signals occupy bandwidth measured in Hertz. In the digital world, signals require bit rates measured in bits per second. But even digital signals have waveforms, which occupy frequency space measured in Hertz.

Telephone speech can be converted to a digital format, usually at a bit rate of 64 kbps. Data is already digital and operates over a wide range of speeds, with personal computer modems operating at 56 kbps and high-speed modems at about 1 Mbps. Television can be converted to a digital format in the range of 1 Mbps to 100 Mbps, depending on compression and quality. Thus, speech, data, and video can all operate in a digital format in which a bit's a bit. To many people, this common digital format represents a convergence of signals. But in the old world of analogue, speech, data, and video all occupy analogue bandwidth in which a Hertz is a Hertz. It thus means little philosophically whether signals are analogue or digital, because the format of a signal is really an engineering question depending on the need for noise immunity.

The telephone network uses a form of switching known as *circuit switching*. The Internet uses a form of switching known as *packet switching*. These two approaches to switching are quite different. Circuit switching seems best suited to voice telephony, although it can also be used for data communication. Packet switching seems best suited to data communication, although it can also be used for voice telephony.

With circuit switching, a continuous connection is maintained for the duration of the communication. This connection can be a real circuit formed by a physical electrical connection between the two parties. This connection can also be a virtual circuit created by transferring digital bits in such a manner to give the appearance of a physical connection. With circuit switching, the communication flows effortlessly and continuously, whether or not any real information is communicated. The connection, be it physical or virtual, is always there and available for use.

With packet switching, information is broken into a series of packets, and each packet is sent separately from the source to the destination. Depending on the specific protocol or standard being used, packets can be fixed in length (either very short, consisting of about 1,000 bits, or very long) or variable in length. In addition to the actual information being sent, each packet also contains header information along with information specifying the addresses of the destination and source.

The historical evidence is that the pace of growth in capacity of transmission systems is accelerating. More recent generations of transmission systems have compounded annual growth rates much faster than older systems. Switching is dependent on computer technology, and computer processing power grows by Moore's law at a constant growth rate. Thus, transmission bandwidth is outpacing switching. This could change if a new generation of switching systems, such as optical switching, were developed. In the meantime, one of the rationales for the development of the Internet and packet switching, namely, to save bandwidth, is today much less valid. Packet switching might therefore be replaced by circuit switching, which wastes bandwidth but has simpler switching and little latency. There is always a trade-off between bandwidth and latency. The real issue today is not bandwidth, but switching.

Most switched networks are organized into a star configuration, with switching occurring at the hub of the star. A bus configuration is used mostly for broadcast applications, such as cable television, in which the same signals are sent to everyone. A bus configuration can also be used for switched data, such as the Ethernet protocol, in which the data signals include information to specify the destination.

Most engineers would consider cable television to be a broadcast, nonswitched architecture. However, switching does occur at the set-top cable box where the viewer chooses which channel to watch. In effect, all channels are sent to everyone, and the viewer chooses which one to watch. This is clearly a form of switching.

## NETWORK REQUIREMENTS FOR SWITCHED VIDEO

Conventional broadcast VHF/UHF television today remains analogue, but direct broadcast satellite television is already compressed digital. Digital TV is also being broadcast over-the-air by the conventional TV stations utilizing additional UHF channels allocated for that purpose. Nearly all TV studios are heavily digital already. Thus, it is not "digital" that characterizes Internet TV. What characterizes Internet TV from conventional television is the use of a packet-switched network to deliver video, either downloaded or streamed in real time. But a number of technological challenges need to be overcome for this switched video to happen. One is the amount of traffic generated by video.

Claims continue to be made that data traffic exceeds voice traffic. But studies of users show that for switched networks voice traffic greatly exceeds data traffic by a factor of at least ten (Noll, 1991; Noll, 1999). Much of the actual data traffic could be search services and overhead, although the downloading of programs and media is an ever increasing trend. Telephone speech, when converted to a digital format, consumes 64 kbps in each direction. Thus, 10 minutes of two-way speech consumes a total of 76.8 million bits. It would require a tremendous amount of Web surfing to consume this amount of bits, and many people spend an hour or more each day talking on the telephone, wired or wireless. All this changes for video.

Two hours of digital video compressed to an average transfer of 4 Mbps consumes a total of 28.8 billion bits. These 2 hours of compressed video require nearly 400 times the capacity required for 10 minutes of a two-way telephone conversation.

Transmission capacity in backbone networks is probably available for Internet television. A single strand of today's single-mode optical fiber routinely carries a few Gbps. The theoretical maximum capacity of a single light-frequency channel of single-mode fiber is 200 Gbps, which is enough capacity to transmit 50,000 compressed TV programs. As impressive as this may be, if the entire light spectrum were used, then the theoretical capacity increases to 50,000 Gbps (Noll, 1998). This is enough capacity to carry 12.5 million compressed video programs at 4 Mbps each. If audiences are bored by today's 200 channels over DBS, then imagine 12 million channels from which to choose!

Intra-premises wiring in the form of coaxial cable or even twisted-pair of copper wire has considerable capacity to carry video and is not a problem. The access from the home to the local access point might appear to be a challenge. The coaxial cable of the CATV provider has considerable capacity, but this capacity is being utilized fully to transmit conventional video programs and there is not sufficient spare capacity to transmit Internet television to hundreds of individual users. Systems in which a single circuit is shared, such as CATV, could be overwhelmed by many simultaneous users. One solution would be to use fiber rather than coaxial cable because fiber has tremendous capacity. However, technologies to enable sharing of fiber by many simultaneous users on the same fiber strand are still costly, although they surely will be developed. Another solution is to dedicate a separate fiber to each home, but this too is a costly architecture to install.

The local telephone network uses a separate twisted pair of copper wire to each home and does not suffer from being overwhelmed by many simultaneous users. The only problem is the capacity of the twisted pair that depends on distance. Over relatively short distances (under a mile), twisted pair can easily carry a conventional analogue television channel. Digital subscriber line (DSL) technology exploits this capacity by placing digital information in frequencies above the 4-kHz voice baseband signal. However, DSL is currently limited to distances less than 15,000 feet. One solution is to bring fiber to the neighborhood and then complete the high speed connection over copper wire. The local access architecture will continue to evolve, and as one bottleneck is resolved, another will appear.

Clearly, there is considerable transmission capacity in today's network, particularly the intra-premises and backbone portions, with plenty of options for local access. The real challenge to Internet television is switching. Internet video pushes packet switching by a factor of 100 beyond Internet access at conventional 56 kbps rates. In addition, because everyone watches television at the same prime time, the requirement for simultaneous access makes it difficult to time share switching facilities. The switching of all the packets required for video would overwhelm today's packet backbone along with local Internet access facilities. It is switching, and not transmission, that would be overwhelmed by Internet television.

Switching is closely related to computers, particularly because digital switches look much like a digital computer. Servers are computers that handle the information needs of many simultaneous users. Servers to handle videos accessed and delivered to thousands of simultaneous users do not exist and will be a challenge to develop. One solution is to distribute many servers throughout the network as close to users as possible so that each server handles an acceptable number of simultaneous viewers. But such a large investment in technology would be costly and would be quickly obsolete as the technology progressed.

## OTHER TECHNOLOGICAL CHALLENGES

The issue of the convergence of the home TV set with the home personal computer is perplexing. Although TV sets increasingly are utilizing computer technology and although personal computers utilize visual displays, TV sets and computers remain much apart. Most consumers do not care about convergence. The TV set is used in a passive manner to watch television and view videos. The personal computer is used to access and send e-mail and to obtain information from various Web sites. As Internet users become more experienced, their web surfing decreases as they discover and bookmark their favorite sites. If television is ever delivered over the Internet, then software reliability will need to be improved greatly. It seems that most personal computers crash a few times a day. This will not be tolerated by most consumers for simply watching television. Internet TV will need to be robust.

One challenge facing Internet television is navigation through thousands of choices to reach the desired video program. This problem occurred with DirecTV and was solved through on-screen program listings to assist the viewer in navigating to a desired program. The use of an Internet-like interface to assist in the navigation through listings of hundreds and thousands of programs would make sense.

Set-top boxes are supplied by CATV providers as an interface between the home TV set and the coaxial cable. These boxes are costly to develop and install on a massive level. The use of the Internet protocol to deliver television over CATV will require a new generation of set-top boxes. Moreover, before such boxes are developed, standards must be agreed on and adopted.

Many standards questions abound with Internet television. Systems must be interoperable, compatible, and reliable. In terms of compression, MPEG is already the "HTML" of video. But Internet television will introduce the need for a host of new standards.

## FREEING UP THE AIRWAVES

Internet television is a switched service that could be delivered over coaxial cable, optical fiber, and twisted pair of copper wire. Thus, Internet TV would continue the trend of moving broadcast television away from the air waves of the VHF/UHF spectrum. Cable television and direct broadcast satellite television are the ways most Americans receive their television. Only about 20% of Americans presently receive their television directly over VHF/UHF (Noll, 1999). If Internet TV were to decrease the use of VHF/UHF television, then interesting policy questions arise concerning whether or not the VHF/UHF channels should be returned to the public by the broadcasters and used for more valuable purposes, such as wireless cellular telecommunication. As policy, the government might then require CATV and Internet TV to offer a minimum "lifeline" television service for free. The former VHF/UHF broadcasters might be required to pay these alternative media to distribute their programs for free. This would be fair because the broadcasters would be saving the expense of over-the-air broadcast transmission.

## SWITCHED VIDEO

Television currently is transmitted over broadcast, nonswitched networks. One definition of Internet television is the transmission of television over the Internet, which is a packet-switched network. Why would television migrate from a broadcast medium to a switched medium? One reason would be that the switched medium was less costly, but the packet-switched Internet is more costly than conventional broadcast media. Another reason would be new features for viewers. But, thus far, all attempts at interactive television (other than teletext in Europe) have met with poor consumer acceptance.

There is one video service that requires a two-way, switched network, namely, the videophone or picturephone. Will Internet TV then evolve into the videophone? This is doubtful, because all evidence indicates that most people would rather not be seen while speaking on the telephone (Noll, 1992). However, many people are using the Internet to send videos of children and family trips to their friends and family. This trend could expand and ultimately evolve into the videophone.

During the early days of the picturephone in the 1970s, there was discussion of the use of cameras at public places that could form a switched video network. But the need for such switched video was not clear back then. Today, web-cams are located at key traffic places, such as bridges and tunnels in the New York City area, to show the current traffic situation. The still images of these web-cams could evolve into full motion, a form of Internet television.

Most people care most about the news that has the most impact and interest to them, and that is mostly about their immediate family and friends. The larger the scope of information, the less the importance to individuals. This hierarchy of the need for information starts at the individual and extends to the universe with family and friends, neighborhood, city and state, and country in between (see Fig. 3.2). Broadcast radio and television do fine at broadcasting news of broad interest, but do nothing for the neighborhood and family news

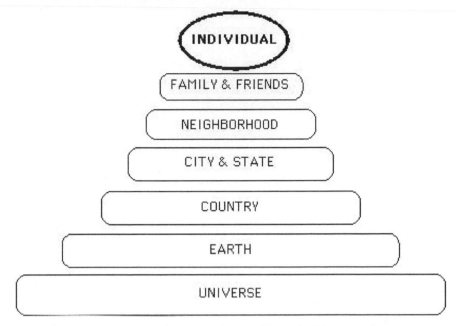

FIG. 3.2.    The need for information forms a hierarchy. Most people are most interested in information that impacts on their immediate family and friends. Information about the Earth and the universe might be interesting, but the need is not very strong or pressing.

that concern people most. Internet TV might have an ability to serve more focused, narrower content of such neighborhood and family news.

## REFERENCES

Noll, M. A. (1991, June). Voice vs. data: An estimate of future broadband traffic. *IEEE Communications Magazine, 29*(6), pp. 22, 24–25, 78.

Noll, M. A. (1992, May/June). Anatomy of a failure: Picturephone revisited. *Telecommunications Policy, 16*(4), pp. 307–316.

Noll, M. A. (1999a, June). Does data traffic exceed voice traffic? *Communications of the ACM, 42*(6), pp. 121–124.

Noll, M. A. (1998). *Introduction to telephones and telephone systems,* (3rd ed., p. 114). Boston: Artech House.

Noll, M. A. (1999b, October). The impending death of over-the-air television. *Info, 1*(5), pp. 389–391.

# Network Business Models and Strategies

# II

## Network Business Models and Strategies

# Industry Structure and Competition Absent Distribution Bottlenecks

Michael L. Katz
*University of California, Berkeley*

Today, television broadcasting and the Internet are at opposite ends of the spectrum along several dimensions. Broadcast television has relatively high production values, limited consumer choice, and is one-to-many. Internet services generally have very low production values, offer tremendous consumer choice, and can be one-to-one, as well as one-to-many. Technological progress has the potential to break down many of these distinctions.

For at least a decade, discussions of convergence in telecommunications have focused on the convergence of voice and data. Over the next decade, convergence will extend to video. This extension may take several forms. It may entail the current broadcast television infrastructure's being used to offer existing Internet services, such as e-mail and web browsing. Alternatively, it may become possible to offer television-quality video over the Internet's wired infrastructure at low cost.

Although both forms of convergence are probable, the focus here is on the second: What is likely to happen when the Internet can be used to carry individually selected, full-motion video programs to the vast majority of the U.S. population at relatively low cost?[1] The rise of entirely new and

---

[1]The focus is on the United States because of its important global roles in both the television and Internet industries. Although other countries often have a very different market structures, many of the market forces identified here are relevant to those countries as well.

unforeseen services will very likely turn out to be the most far-reaching and important development. However, it is hard to say much about unforeseen services. Thus, the predictions offered are restricted to services closely related to today's television and the ways in which Internet distribution may affect the industrial organization of, and competition in, the television industry.

This chapter first offers a set of criteria by which to determine what constitutes television for present purposes. It then identifies four important technological trends on which the analysis of television is predicated. Next, the analysis of market effects begins by briefly addressing questions concerning the fundamental business models for television in the light of these technological trends. The next section presents a decomposition of the current television value chain into separate stages and examines how competition in each stage is likely to be affected by the projected technological trends. The implications of these technological trends for the extent of vertical integration and bundling across stages of the value chain are then examined. Lastly, some thoughts are offered on how the predicted developments will affect the economic welfare of existing industry participants. Consumers are likely to be the big winners, and local broadcasters will be the big losers from Internet distribution of television. The effects on other parties will generally depend on their abilities to take advantage of new opportunities. Brief thoughts on the likely timing of various developments are offered in a concluding section.

## HOW WILL WE KNOW "TELEVISION" WHEN WE SEE IT?

In order to answer questions about the future of television, a definition of *television* is required. Television can be defined in terms of content (e.g., video news and entertainment), a transmission technology (e.g., wireless transmission within a particular bandwidth), a form factor for receivers (e.g., specifications for TV screens and how close one sits to them for viewing), or even a social context (e.g., whether viewed in a group or alone).

The rise of the Internet to distribute full-motion video is likely to change people's conception of television itself, as well as the industry that provides it. In order to stay focused, this chapter concentrates on services that are much like what is seen on television today. In particular, it adopts a loose definition with the following elements.[2] The flow of content is asymmetrical; the bulk of the information flow comprises "programs" sent from the "service provider" to the "viewer." Messages from the viewer principally consist of instructions to the service provider. The programming is

---

[2]This definition is similar to one offered by de Vos (2000, p. 13): "In principle, television is the public transmission, over some distance, of audiovisual programmes and services made for a relatively large audience."

created by professionals for relatively large audiences. The programming comes in discrete units of between 15 minutes and several hours. Lastly, a viewer can sample and select the programming relatively quickly, if not instantaneously. Whereas transmission need not be instantaneous or real time, it must be "convenient time." Whether "convenient" turns out to mean a few seconds or a few minutes remains to be seen.

As illustrated in Figure 4.1, television will be only a small part of the forthcoming multimedia Internet.[3] Moreover, there will be few bright-line boundaries between services. Nevertheless, the criteria above help distinguish television from a number of other video services that might be offered over the Internet:

- E-videotape can be thought of as services that require planning to utilize (say an hour or two before viewing or even a day in advance to allow overnight downloading). Using such a service would be similar to a trip to a video store, without the inconvenience of travel and with a much larger, constantly updated selection.[4] Al-

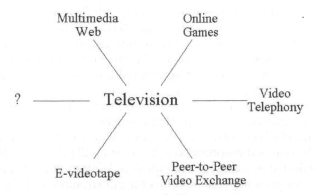

FIG. 4.1.   Television within a broader web of applications.

---

[3]Indeed, there is reason to believe that television services as they have been defined will not drive the deployment of high-speed access to the home. Broadcast and cable television will continue to offer programming and advertising at relatively low cost during the time frame over which broadband is deployed. Consumers' incremental willingness to pay for increased variety and interactivity of television programming is uncertain at this time.

[4]Of course, there is the ongoing debate of whether there are desirable aspects of the video store experience (e.g., buying a box of Milk Duds for immediate consumption or looking for other single people renting movies) that cannot easily be replicated through electronic media.

though such a service may be a relatively close substitute for television for viewing dramas, comedies, and documentaries, it would not be a good substitute for the delivery of fast-breaking news or major sporting events. Such a service could also place different demands on the network and edge devices than will television, substituting storage and processing capacity at the consumer's premises for network transmission capacity.

- Another type of service is peer-to-peer or server-mediated end-user video file exchange (e.g., a "video Napster"). Video file exchange is similar to e-videotape when the files exchanged are professionally produced material, as opposed to something more like video mail.
- Video telephony is a real-time, symmetrical, two-way service that is generally one-to-one without professionally produced content. The way in which consumers use the service will be different from television, as will the demands the service places on the network (e.g., very low latency and bidirectional capacity).
- Online video games have asymmetrical information flows and have professionally produced content, but the degree of interaction is still recognizably greater than television as defined here. There will be some blending of online video games and television as television programs continue to add interactive elements, such as playing along at home with the on-screen game show contestants or guessing the next play called in a televised football game. But these applications demand considerably less of the network in terms of latency than do action video games. Television programming can rely on streaming and the use of buffers to allow transmission over less capable networks.
- Multimedia web services with short video clips integrated into largely text- and graphics-based content will primarily differ from television in terms of user interaction. With multimedia web services, the user actively searches for information and digests it in relatively small chunks. Multimedia web services also place weaker demands on the network in terms of capacity.
- Lastly, entirely new services may develop.

## CENTRAL TECHNOLOGICAL TRENDS

Technological progress in computing and telecommunications is giving rise to at least four developments that have fundamental implications for business models and competition in television.[5]

---

[5]There is a fifth technological development that has important policy implications: Global connectivity of the Internet raises jurisdictional issues for content regulation.

## Increased Ability to Process User Feedback

Perhaps the biggest change is the development of a return channel that allows the viewer to send information all the way back to the intelligence in the program provider's system. The existence of back channels carrying messages from the viewer to the service provider creates possibilities for several new types of services. One possibility is for the service provider to pass the end-user's message on to a third party, either to exchange information (e.g., e-mail) or to facilitate some form of e-commerce transaction.

Another possibility is to use the back channel for the viewer to communicate with the video service provider itself. One use of this capability is for the service provider to collect information about the viewer that can then be relayed to advertisers. This back channel can also be used to allow the service provider and end-user to customize the content that the end-user views. For example, an end-user may choose among camera angles when viewing a sporting event. Or she may choose the language spoken by the commentators. Consumers enjoy a degree of interactivity today (e.g., they choose the channel to watch), but the set of choices is constrained by the limited capacity current distribution networks. The future ability to offer customized programming will be a consequence of a broader effect of the back channel—the ability to tailor the signal that is sent to a particular end user will create a dramatic increase in *effective* distribution capacity. This point is sufficiently important to break it out separately.

## A Tremendous Increase in Effective Distribution Capacity

Existing over-the-air broadcast television provides distribution at a very low cost, but offers relatively low capacity in terms of the variety of programming.[6] Cable and satellite distribution systems greatly increase the number of channels that can be broadcast, but transmission capacity still is tiny compared to the collection of available programming. One reason for the lack of capacity is that existing distribution technologies send signals for all programs to all consumers and then filter out the unwanted ones at the viewers' premises. In a sense, there exists a very short back channel from the viewer to the program provider that reaches only as far as the television set or set-top box. This situation is illustrated in the top panel of Fig. 4.2. The figure illustrates the fact that a wide range of content is filtered down for broadcast and then further filtered when the viewer selects a particular channel. As shown in the bottom panel of Fig. 4.2, extension of the back channel and creation of user-dedicated transmission channels (i.e., switched broadband access) change the situation com-

---

[6]In making this statement, I—like the U.S. Congress—am ignoring the tremendous opportunity cost of the spectrum.

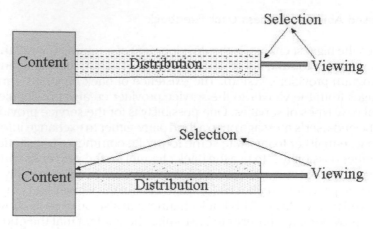

FIG. 4.2.   Lengthening the back channel.

pletely.[7] The dedicated channel can be used to carry any properly format-ted program in the content library that is selected by a connected viewer.

### The Separation of Applications from Transport

The layering model of the Internet allows for the development of applica-tions that are oblivious to the underlying transport infrastructure. This pattern is sometimes referred to as the hourglass structure of the Internet architecture because there are minimal specifications of protocols in the middle that support a wide range of transport networks below and a wide range of applications above (Computer Science and Telecommuni-cations Board, 2000). (Figure 4.3 illustrates this structure.) This architec-

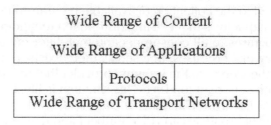

FIG. 4.3.   Hourglass architecture.

---

[7]Although they are not explored here, this increase in capacity has important policy implications. Much of the current U.S. broadcast regulation is an ostensible response to spectrum "scarcity." The already tortured arguments for much of this regulation should be strained to the breaking point.

ture allows innovation to occur at the applications layer and the transport layers separately. The feasibility of independent innovation speeds the rate of innovation and increases the flexibility of the network to take advantage of new opportunities. Someone with an idea for a new application can bring it to market without having to alter the underlying transport infrastructure. As discussed in greater detail later, this technological separation also facilitates the ownership separation of distributors from content creators and packagers.

### Continued Increase in Storage and Processing Power Controlled by Viewers

Another (and related) characteristic of the Internet is that the intelligence resides at the edges of the network. Internet devices typically are "smart," in comparison with "dumb" televisions, suggesting that Internet televisions will be smarter than current models. Indeed, consumer television devices already are getting smarter with advent of products such as TiVo, which allows a viewer to record programs on a hard drive and then manipulate the data in various ways, such as pausing. As users have increasing processing power and memory under their control, they can engage in editing, time shifting, and copying, among other activities. This type of user control over programming potentially has profound implications for business models, as the next section discusses.

### WHO IS GOING TO PAY FOR "TELEVISION" IN THE FUTURE?

Most suppliers in the television industry are there to make money. Television content creators, aggregators, and distributors' revenues ultimately derive either from payments made by advertisers or from subscription fees paid by viewers.[8] Certain technological developments threaten business models based on either the subscription or advertising revenue streams. There are, however, several possible supplier responses. Moreover, other technological developments may strengthen these business models.

Consumers' ability to copy programming (and the development of widely deployed peer-to-peer communications) threaten suppliers' ability to rely on subscription fees. Service providers can be expected to implement various forms of copy protection in response. However, history suggests that these measures will be defeated, if by nothing else then by consumers' conducting video screen scrapes. Interestingly, the lengthened back channel and the ability to offer interactivity can create personalization that may make copying more difficult and costly. With interactive

---

[8]For a more detailed description of the current industry structure, see Owen and Wildman (1992).

programming, viewers may only see the results of a particular interaction, not the underlying program that drives the creation of each instance. Although it falls outside of the present definition of television, consider a video game played over the Internet. Even if a player could readily copy all of the images on his or her screen, sending those copies to another person would be a poor substitute for a copy of the game itself.

Consumers' ability to edit programming affects suppliers' ability to rely on the sale of advertising. Increases in memory and processing power will make it increasingly easy for consumers to avoid commercials.[9] However, several strategies will be available to service suppliers to counter this trend. One is to create commercials that consumers want to view because the advertisements are entertaining or informative. A second strategy is to provide a separate reward to consumers for watching advertisements they would otherwise like to avoid viewing. Consumers' viewing of commercials would be monitored (e.g., consumers might respond via the back channel to instructions or questions embedded in the advertisements), and consumers would be rewarded for watching commercials by being given monetary payments or conditional access to desirable programming.

Another possible strategy in response to consumer sophistication is to create advertisements that cannot be avoided because they are embedded in programming consumers desire to watch. The suggestion that television is moving to a world of ubiquitous product placements is only somewhat facetious.

Whereas some technological developments threaten the advertising business model, others will create new opportunities. Digital technologies will create enhanced product placement capabilities. For instance, technology makes it possible to combine multiple signals on a single screen in an integrated fashion, and different consumers may well see different products in the same place on their screens. This fact raises questions about who will control the screen a viewer sees. Who will control banner ads, electronic product placements, and other forms of advertising or electronic interaction on the screen? The answers to these questions will have profound implications for business models.

New technologies will also make it possible to offer advertisers better monitoring of viewing patterns and more tightly focused viewer demographics. The latter can be attained in two ways. First, the increased fragmentation of viewing audiences due to the creation of targeted programming (discussed later) will offer advertisers narrower audiences

---

[9]Viewers have some tools at their disposal today. Those viewers watching a stored (on tape or a disc) copy of television program can fast forward through the advertisements. This process may become automated. On the World Wide Web, AdSubtract (http://wwww.adsubtract.com) already offers software that blocks banner ads on web pages.

along the lines that cable television has done. Second, even when a program has an audience with diffuse demographics, it will be possible to transmit different advertisements to different viewers for finer segmentation than that provided by audience self-selection alone.

In any event, the remainder of this chapter assumes that the technologies necessary to support both the advertising and subscription models will be created.[10]

## IMPLICATIONS OF TECHNOLOGICAL CHANGE FOR THE VALUE CHAIN

Assuming that a successful business model is developed, how will the rise of the Internet as a medium for distributing video content affect television? The analysis first examines the value chain for video programming and the ways in which the technological developments already discussed will affect *individual* links in the value chain. The following section takes up the issue of how Internet distribution affects relations *across* links of the value chain.

### The Current Television Value Chain

Figure 4.4 lays out a simple value chain for video programming. This value chain does not explicitly illustrate the production and sale of advertisements, even though advertising is the primary output of the over-the-air television industry. This value chain is nevertheless useful because viewers provide the eyeballs for which advertisers are paying, viewers pay subscription fees, and viewers will provide a customer base for future e-commerce transactions.

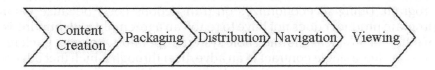

FIG. 4.4.   A simple value chain for video programming.

---

[10]There is a slight danger in this approach in that an entirely new business model might develop that could have profound effects on the industry structure. This is not likely to happen, and in any event who knows what it will be.

### Content Creation

Content creation consists of the various activities undertaken to produce the programming ultimately offered to viewers. Television program creation is undertaken by major studios, as well as a variety of independent producers. Local broadcasters also create content, primarily in the form of various news, and what might be called news-lite, shows.

### Packaging

There are several dimensions to packaging.

*Filtering.*   Even with cable and satellite transmission, the ability to distribute programming to viewers is greatly limited relative to the potential demand for program variety. There are tens of thousands of programs and hundreds of millions of potential viewers who may want to watch different programs or view the same program but at different times. Hence, an important role today is to select which programs are broadcast and which are not broadcast. Local broadcasters, satellite broadcasters, cable systems operators, and broadcast and cable networks all play this role.

*Timing.*   Timing involves strategies based on the relations of programs within and across channels.[11] Within channels, packagers worry about flow, or how the audience of a program feeds into the audiences of programs following it. Packagers also worry about what program to show against programs on rival channels at a given time. Today, local broadcasters and broadcasting and cable networks are the primary timing decision makers.

*Aggregation.*   At present, packagers engage in several types of aggregation. Terrestrial over-the-air broadcasting and cable television systems distribute video programming over local areas. Packagers engage in geographic aggregation in response. Networks and syndicators reach contracts with a large number of local outlets to broadcast a given program. Doing so economizes on transactions costs because each program producer does not have to reach an agreement with hundreds of different broadcasters and cable systems operators. Second, packagers can offer a single contract to an advertiser through which that advertiser can purchase advertising time on a large number of local distribution systems at once. A network can also provide cross-program aggregation that offers one-stop shopping to a potential advertiser seeking time slots on a range of programs. Lastly, satellite and cable systems

---

[11]Although timing can be conceptualized as a particular instance of filtering, it useful to discuss the two concepts separately.

operators aggregate consumers in collecting subscription fees for networks, again economizing on transactions costs.

### Distribution

Distribution consists of delivering a signal that carries content from packagers to television receivers located on potential viewers' premises. Today, the dominant forms of distribution are terrestrial over-the-air broadcasting, cable television, and direct-to-home satellite broadcasting.

### Navigation

Navigation services have at least two components. One is simply to tell potential viewers when and on what channel programming is available. Today, such services are provided by on-screen guides, on-air promotions, newspapers, magazines, and Web sites. A second component is to provide various ratings or advice to potential viewers. Newspapers, magazines, and Web sites all offer opinions on the quality of various programming.

Clearly, the activities of the navigation stage are related to the filtering of the packaging stage. When multiple programs are offered as a package under a common brand, the brand may develop a reputation that consumers use as a basis for making viewing (or at least sampling) choices. These reputations can form as consumers make predictions of the likely quality of unseen shows based on the consumers' experiences viewing other shows offered under the same brand (e.g., on the same network). Perhaps the central difference is the extent to which the process is one of narrowing down the set of choices offered to a consumer (packaging) versus helping the consumer make choices from a wide universe (navigation).

## Changes

The distribution of full-motion video over the Internet to a mass audience will dramatically affect the distribution link of the value chain. This development will then have significant effects on the other stages of the value chain.

### Distribution

Because they are central to overall developments, consider first changes in the distribution stage. Internet distribution of television entails moving bits from packagers' servers to computers located on viewers' premises over the various networks that make up the Internet infrastructure. With the advent of widely deployed broadband video services, there will no longer be a *technological* bottleneck in distribution. It remains to be seen whether there will be a *commercial* bottleneck due to monopoly or duopoly control of local (or "last mile") broadband distribution.[12] At least for traditional programming, as long as satellite and terrestrial over-the-air broadcasters remain, the commercial bottleneck will not be any narrower than it is today. Even if there remains a commercial bottleneck, it need not limit the variety of programming made available to consumers. Whether the bottleneck limits variety will depend both on the business model adopted by any bottleneck distributor as well as public policy toward that distributor. This issue is addressed in a later section.

### Packaging

Changes in the distribution of television will have major impacts on all three of the roles that packagers play.

*Filtering.* The increase in the length of the back channel has the effect of making all programming available to potential viewers once the content has been stored in digital form. There is no longer a demand for filtering in response to capacity constraints in the distribution system. Instead, the scarce resource is viewer time (and money). Thus, the filtering function will shift almost entirely to the navigation link in the value chain.

*Timing.* The change in the distribution model increases the possibility of widespread asynchronous viewing. In an asynchronous world, the role of the packagers in creating flow and engaging in counterprogramming may be greatly diminished. Instead, viewers may work with navigation services that allow viewers to create their own packages for many types of programming.

Synchronous viewing will not disappear. Presumably synchronous viewing will remain important for major sporting contests, awards shows, and other forms of event programming. Synchronous viewing may well continue for comedy and drama series because of benefits of common viewing times that allow people to discuss particular programs with their friends and coworkers the next day.

*Aggregation.* As many observers have noted, once a site is connected to the Internet, it is globally available. This fact suggests that the explicit geographic aggregation role of packagers will disappear. Indeed, there may be a role for geographic *dis*aggregation, whereby service providers offer targeted advertising or institute charges for advertising based on the locations and other demographics of viewers.

---

[12]At least for very densely populated areas, a degree of oligopolistic competition is probable. Faulhaber and Hogendorn (2000) presented a calibrated simulation indicating that 70% of households will have a choice among at least three providers if there is a 66% take rate for broadband access at $50 per month.

Other types of aggregation may become more important than at present. For instance, advertisers place value on being able to reach a mass audience with a single campaign. One way to accomplish this objective is to run an advertisement on a program with a very large audience. The increase in distribution capacity and the resulting number of programs available to viewers will fragment audiences even more than have cable and satellite distribution, making this strategy increasingly difficult. An alternative approach is to show an advertisement in a coordinated fashion across a large number of programs simultaneously. By aggregating a large number of smaller audiences, a packager can thus create a synthetic mass audience.

Programs may also be aggregated in the sense that a variety of programs are offered at a single site or under a single brand name. As discussed earlier, branding can allow firms to form reputations and thus serve as a form of quality certification on which consumers could base viewing choices. In this way, the packaging role would shift to become a navigation role.

Lastly, program providers whose business models rely on payments by viewers may also pursue packaging or bundling strategies. Two extremes frame the possibilities. Under total unbundling, programs would be available in small units (one could imagine charging by seconds of viewing time). In this world, packaging would be limited to branding and the creation of stand-alone dramatic units of programming, and navigation would play a very large role. At the other extreme, programming would be combined into a handful of large bundles and sold to potential viewers only as packages.

It appears that the industry will continue to offer a mix. The continued evolution of the online payments industry will reduce the costs of offering pay-per-view programming over the Web. In addition, the increase in distribution capacity will eliminate the need for channels that provide little bits of all types of programming (e.g., news, sports, comedy, and dramas) as do the traditional broadcast networks. Instead, viewers will be able to select among a huge menu of specialized offerings—taking the development of specialized channels on cable television several steps further. Nevertheless, service providers may charge fees for bundles of programs, rather than on a program-by-program basis, as a strategy to extract surplus from consumers.[13]

If much of this sounds somewhat familiar, it should. These predictions regarding the role of packagers for television over the Internet mirror the

---

[13]Horizontal bundling strategies may allow firms to engage in certain forms of price discrimination. Moreover, horizontal bundling strategies can affect the nature of competition between service providers. Nalebuff (2000) shows that a firm can gain strategic advantage by selling a bundle in competition with a collection of firms selling individual components.

role of packagers of text and still images on the World Wide Web today. In terms of overall structure, the World Wide Web has a similar value chain to television. And the similarities are likely to increase as television moves to the Internet. Despite the possibility of offering virtually unlimited variety on a per-program basis, packaging takes place (in the form of branded Web sites offering bundles of information, in some cases for a fee) and coexists with extensive independent navigation services (e.g., Yahoo!) and various search engines (e.g., Google).

## Navigation

The increase in both the number of programs and the sources of programs will greatly increase the need for both navigation-as-a-map and navigation-as-an-advisor. Viewers will be looking for comprehensive program guides that provide good predictions of whether they will value various programs. Given the limited number of channels available to many viewers today, viewers can attempt to sample (i.e., channel surf) to determine what programming to view. In the future, doing so will be nearly impossible; imagine randomly wading through the millions of sites on the World Wide Web without a search engine or directory to find interesting pages. Web sites will certainly develop that provide recommendations and reviews of Internet television programming, as well as offer search engine capabilities. Presumably, there will be specialized search engines appealing to particular tastes and search engines that build on a user's viewing experiences to refine future searches.

The relatively low costs of setting up such sites, coupled with heterogeneous viewer preferences and the possibility of creating targeted sites, should lead to a monopolistically competitive or oligopolistic market for navigation. That said, the scalability of navigation technology may lead to guides with high market shares that come in many versions, perhaps even tailored to the viewing history and tastes of each person individually.

## Content Creation

Given the ability to distribute a wide range of programming, it should not be surprising to see several different developments simultaneously. These developments will have a common thread: There will be increased competition to attract viewers and thus there will be demand for programming that is increasingly attractive to viewers. The increase in distribution capacity provides the opportunity to offer shows highly valued by relatively small numbers of potential viewers. Thus, niche programming targeted at particular viewers' interests will be offered. Just as cable offers more specialized programming than does broadcast television, the Internet will offer more specialized sites than does cable. The specialization may have a geographic component. In Europe, for example, increases in the number of broadcast channels led to more programming of local interest (so-called proximity TV) (de Moragas Spa, Garitaonandia, & Lopez, 1999).

At the same time, the emergence of a potentially seamless global distribution mechanism will increase the rewards to programs that have broad, international appeal.[14] Thus, there may be huge expenditures on high-end, mass appeal programming, similar to the present motion picture industry. There is reason to expect that most viewing will be of a relatively small number of programs, as one observes with television today.[15]

Just as video games and broadcast television coexist today, in the future there will almost certainly be programming with a wide range of interactivity. In fact, a given program may offer viewers a range of degrees to which they are interactively involved.

## IMPLICATIONS FOR VERTICAL STRUCTURE

Having looked at the effects of technological trends on individual stages in the value chain, now consider how these trends will affect the relation between stages. In particular, consider the degree to which vertical integration and bundling are desirable from commercial and public interest perspectives. A firm is *vertically integrated* when it operates in two or more stages of the value chain. A firm engages in *vertical bundling* when it makes its services at one stage available in only fixed combinations with services at another stage.

### The Current Extent of Vertical Integration and Bundling

At present, many industry participants are vertically integrated into two or more stages of the value chain. Although primarily packagers, the broadcast networks generally are backward integrated into content creation and forward integrated into distribution. All of the major broadcast television networks have in-house production arms for television programming, and many networks are associated with major motion picture studios. The parents of the major broadcast networks tend to be the larg-

---

[14]"Potentially" seamless because there are significant business issues with respect to developing advertising models that will work in a global context (e.g., ads with global appeal vs. location-specific ads inserted based on the viewer's address) and collecting subscription fees on an international basis.

[15]Cable and satellite service subscribers have access to dozens, and sometimes hundreds, of channels. Although most of these subscribers' viewing is of cable networks, their viewing is disproportionately concentrated on programming generated by a handful of broadcast television networks. Despite having fallen for decades, the ABC, CBS, and NBC networks' combined television viewing share is between 30% and 40%. (Paul Kagan Associates, *Cable TV Advertising,* February 28, 1999, and June 21, 1999.)

est group owners of local broadcast stations. And Fox Television's parent, NewsCorp, has financial interests in *TV Guide* magazine and on-screen programming guides, which are navigation tools (News Corporation Website, 2002).

The networks are not alone in vertically integrating. Local broadcast stations focus on distribution but, rather than serve as common carriers, they integrate backward into both packaging and content creation (e.g., production of local news and sports programming). Cable systems operators also engage in packaging by choosing which cable networks to carry, and some large cable systems operators have made significant investments in cable programming networks.

Whereas vertical integration is extensive, there are many firms that operate at only one stage or are vertically integrated but operate as at least somewhat open systems. There is partial but significant unbundling across every stage in the value chain. Despite being integrated into packaging and content creation, broadcast networks buy programming from studios associated with other networks. Networks also purchase programming from independent content creators. Many network affiliates are independently owned, and they typically buy programming from nonnetwork packagers as well as the networks. Cable systems carry cable networks not owned by the systems operators. Lastly, navigation is provided by independent entities, as well as by broadcasters and cable systems operators.

### Potential Benefits and Costs of Vertical Integration and Bundling

A vertically integrated firm may make its services at each stage available separately from one another. Hence, one should consider separately the arguments for vertical integration and vertical bundling. This part presents a tentative assessment of the desirability of vertical integration and bundling in the television industry of the future from social or private perspectives. The central hypothesis is that television over the Internet will be most successful when provided on an unbundled basis on open platforms and that the benefits of extensive vertical integration are limited. Of course, a number of industry participants appear to take a different view.

Proponents of vertical integration ascribe several benefits to it, which stem from the claimed differences in two separate companies' abilities and incentives to cooperate in terms of pricing and investments in comparison with the abilities and incentives of two divisions within a single company.[16] The following are summaries of arguments made in favor of vertical integration, as well as assessments of their probable importance in the television industry.

---

[16]For a summary of arguments for vertical integration that take the view that integration aligns incentives, see Perry (1989).

*Vertical Integration Prevents Double Marginalization.* One benefit ascribed to vertical integration is that it can lead to lower prices when suppliers have significant market power.[17] To illustrate why, consider the incentives of a monopoly supplier of broadband Internet access to raise its price from a set starting point. If that firm is also the monopoly supplier of programming (i.e., is vertically integrated), then it will take foregone programming sales into account when assessing the profitability of an increase in the price of broadband access. But if the programming is sold by a different firm, the access monopolist will not count lost programming sales as a cost and thus has less incentive to restrain price. A similar logic applies to program pricing. This line of reasoning indicates that the sum of the broadband access and programming prices set by an integrated monopolist will be lower than the sum of those prices when set independently by two distinct firms.

The double marginalization logic relies on the existence of suppliers at two or more stages with significant market power. Thus, the problem is considerably reduced if there is competition in the supply of the services at one or both of the two stages. Experience to date suggests that content creation, packaging, and navigation can be supplied by many providers, which should limit their market power. The future degree of competition is more suspect at the distribution level, but the inefficient exercise of market power at this single stage would likely remain a problem regardless of vertical integration or bundling.

*Vertical Integration Increases Investment by Internalizing Pecuniary Externalities.* The separate ownership of different stages in the supply chain can also have negative effects on investment incentives. An investment at one stage may generate benefits for suppliers at a different stage. When the potential investor ignores the benefits created for other providers, it tends to invest too little from the perspective of maximizing the sum of the profits of the two stages. Moreover, the empirical literature on the economics of innovation has generally found that a firm's private incentives to innovate are lower than is socially optimal.[18]

The problem of underinvestment is particularly strong if an investment at one stage induces providers at another stage to raise their prices to appropriate some of the benefits of the investment; the price increase harms the innovator and thus lowers that firm's incentives to undertake the investment in the first place. This effect is an instance of what is known as the *hold up problem*, because once the first firm has made a sunk investment, the other firm is able to "hold up" the investor and appropriate some of the returns to the investment.

---

[17]The problem of double marginalization was recognized by Cournot (1838).
[18]See, e.g., Griliches (1992) and Jones and Williams (1998).

The ability of a distributor with market power to engage in hold up depends in part on its ability to charge different prices for distributing different programs. Program-specific distribution fees provide greater scope for appropriating returns from investments made by any given content creator or packager and thus can weaken the incentives of independent firms to invest in creating content that viewers highly value. Under the traditional broadcast and cable distribution model, a distributor purchases the rights to show content and then charge advertisers and viewers as the distributor sees fit. Under a common carrier model, consumers purchase distribution in the form of transport and then purchase specific programming separately. Thus, a common carrier model of Internet television distribution would be less susceptible to hold up than would the traditional model.

To the extent that there is a potential holdup problem, companies at different stages may recognize the problem. Even self-interested providers of complements can have incentives to cooperate with one another to increase their joint profits. One way is through contracts reached prior to the making of relationship-specific, sunk investments by content creators, packagers, and distributors.[19] Another way is for a firm with market power to encourage investment by developing a reputation for not exploiting its position to expropriate the full returns of investments made at other stages.

*Vertical Integration May Improve Investment Coordination.* In addition to investing too little, independent firms operating at different stages may have difficulties coordinating the nature or direction of their investments. However, industry-wide standards today limit the need for tight coordination between the distribution stage and the content and packaging stages. The layering of the Internet architecture will similarly minimize the need for cross-layer coordination if this architecture is extended to television. Of course, even with continued layering between applications and underlying transport, some types of programming or interactive capabilities may require specialized terminal devices. It has been suggested that there is a need for integration of content producers and customer equipment manufacturers for this reason. But one might reasonably ask whether arm's length cooperation would provide more flexibility and allow firms to specialize in those areas in which they possess distinctive competencies. Content-equipment coordination is, after all, the theory

---

[19]In this regard, it is worth noting that government policy should be careful not to create rules that needlessly limit private parties' abilities to design contracts. Several of the network affiliate rules promulgated by the Federal Communications Commission have this effect and thus create private incentives to integrate. Indeed, there are social incentives for network-station integration because of the inefficiencies that arise when arm's length contracting between networks and their affiliates is limited by both explicit governmental policies and implicit political pressures.

that underlays Sony's disastrous vertical integration into the production of theatrical motion pictures in support of its consumer electronics business.

The previous discussion suggests that the social and private benefits of vertical integration will be limited when television is delivered over the Internet. Moreover, whereas proponents of vertical integration claim the aforementioned benefits, both business decision makers and economists have at best an incomplete understanding of vertical integration. In practice, vertical integration does not necessarily solve the problems identified earlier. Most of these problems arise because actions taken by one firm affect the profits of other and these effects may not be taken into account by an unintegrated decision maker. However, as many people who have worked in or studied large organizations know, different divisions of an integrated firm often are in conflict with one another. Indeed, divisions of companies sometimes get along worse with each other than with outside customers and suppliers. Hence, it is far from evident that vertical integration solves the problems identified earlier or does so better than alternative mechanisms.

Further, vertical integration may have social and private costs as well as benefits. Integration may distort the decisions made by the integrated divisions due to shifts in the decision-making locus.[20] Resources may be wasted on internal corporate politics (e.g., one division may attempt to force another division to rely solely on an input produced by the first division even though the input is substandard), which can be less efficient than the market. Additionally, vertical integration may provide a firm with an increased ability to engage in vertical squeezes that can appropriate the profits of unintegrated rivals and thus undermine the rivals' incentives to make product or process investments (Farrell & Katz, 2000). Lastly, vertical integration appears to create at least some pressures for vertical bundling, which can give rise to social costs. All of these factors suggest that the case for extensive vertical integration is a weak one.[21]

Now, consider vertical bundling. It is useful to frame the discussion of vertical bundling in terms of the costs and benefits of *un*bundling. There are at least three significant social benefits of vertical unbundling.

### Vertical Unbundling Allows the Realization of Mix-and-Match Benefits.
A consumer can take the best offering at one stage and combine it with the best offering at a second stage, even if the offerings are provided by different firms. With Internet distribution, these benefits are potentially much larger than today. There is a huge variety of potential programming, and consumers have widely differing tastes. Internet distribution will create

---

[20]For a theoretical treatment, see Grossman and Hart (1986).

[21]This conclusion assumes that government policies do not unduly restrict private parties' abilities to write contracts that facilitate coordination across vertical stages.

the technological possibility of distributing a much greater variety of programming to satisfy viewer wants. No one distributor or packager will be likely to have all of the desired content. With multiple providers at each stage, consumers would benefit from being able to combine the best match at each level. Notice that this benefit arises even if firms in the industry are vertically integrated as long as they unbundle. Moreover, this effect is both a social and commercial benefit (it improves the gross benefits suppliers can offer consumers).

*Vertical Unbundling Facilitates Innovation by Allowing Single-Stage Innovation.*    Benefits arise when unbundling makes it feasible for a firm that is not vertically integrated to compete by innovating at a single stage or unbundling allows an integrated firm to combine its innovative service at one stage with the services provided by different firms at other stages. The increased creation and diffusion of innovations can be expected to be social benefits. And, to the extent they improve the value proposition that firms can offer to consumers, they are commercial benefits. Broadband distribution and the tremendous potential for innovation over layered platforms will increase the potential benefits of single-stage innovation.

*Vertical Unbundling Reduces Industry Concentration.*    Vertical unbundling increases competition by preventing the most concentrated stage in the value chain from driving concentration in all of the stages, which is what would happen if all firms had to be vertically integrated to compete. Moreover, vertical unbundling facilitates entry by allowing single-stage entry, which reduces the sunk costs (and thus risk) of entry and lessens the need to acquire multiple skill sets in comparison with multi-stage entry. The unbundling of distribution from other stages will thus prevent concentration of the distribution stage from limiting competition in other stages, as Fig. 4.5 illustrates. This pattern is what one has seen in cable television, where the networks carried on a given system are not limited to those owned by the system's operator. The increase in competition gives rise to social benefits by promoting efficiency and consumer welfare.[22] However, from the perspective of incumbent suppliers, it is a "cost."

Vertical unbundling can have social, as well as private, costs. In particular, the following arguments have been made against vertical unbundling.

---

[22]It is sometimes argued that increased competition will reduce innovation. This logic depends critically on what brings about the increase in competition. I am unaware of any evidence that a reduction in entry barriers (other than a weakening of intellectual property rights) has harmed innovation in any telecommunications market. Indeed, in his conference presentation, Dr. Robert Pepper offered data suggesting that competition spurred, rather than discouraged, investment. (Dr. Robert Pepper, "TV Over the Internet: IPTV and Policies for Convergence," conference presentation, Columbia Institute for Tele-Information, November 10, 2000.)

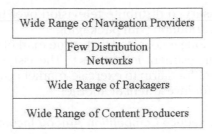

FIG. 4.5. The other hourglass.

*Vertical Unbundling Undermines the Coordination Benefits of Vertical Integration.* One argument is that vertical bundling is needed to realize the potential gains of vertical integration identified earlier. However, a vertically integrated firm can price its unbundled products to take into account the effects on its integrated profits.[23] Similarly, the firm can make investment decisions with the effects on all of its unbundled products in mind. Moreover, it should be taken into account that decreased concentration due to unbundling may create competitive conditions that limit the extent of double marginalization and coordination problems.

*Vertical Unbundling Leads to Underinvestment by Preventing Sufficient Exercise of Market Power in the "Right" Market.* As already discussed, vertical unbundling may lead to increased competition at one or more stages. This increased competition may reduce the ability of firms to extract rents from consumers by elevating prices across the board or engaging in more sophisticated, price-discrimination schemes.[24] It is sometimes argued that a firm with a monopoly in one stage should be able to engage in vertical bundling with another stage because otherwise the loss of potential profits from monopolization of the second stage will undermine investment incentives in the initial stage. This argument has been raised, for instance, in the debate over whether cable companies should have to provide open access when their systems provide broadband Internet access. And it may well arise in the future as broadband transport providers assert that they will not make the

---

[23]The internalization may be less complete, however, than if the firm engaged in bundling. This point is illustrated in the later discussion of bundling's effects on competition.

[24]See Katz (1989) for a review of the economics literature of why a firm at one level might want to integrate downstream to limit competition and support price-discrimination strategies that might not be feasible with downstream competition.

investments necessary to distribute television over the Internet unless they can bundle distribution with packaging.

Public policy has long recognized that some market power can be necessary to provide investment incentives.[25] The issue here, however, is whether it is efficient for a firm to exercise market power at one stage to provide incentives for investment at another.

### Unbundling Can Reduce Competition Among Incumbents.

Unbundling alters competition among a given set of incumbents in ways that can reduce suppliers' incentives to price near costs. Consider, for example, a situation in which there are only two stages and there are two firms, each of which produces component services at each stage. Suppose that each of the firms is a lower cost producer of one of the two components. When the firms sell their components individually, the lower cost producer of each component sets its price just below that of the other supplier. Hence, consumers pay a total amount for services equal to the sum of the higher costs of each component. When the firms compete by offering bundles, the firm with lower average costs of the two components wins sales at a price just below the other firm's average cost of the two components. This firm's average cost of the two components is manifestly less than the sum of the two higher costs of each component across firms.[26] Intuitively, bundling leads to lower prices because a firm is willing to "take a loss" on its high-cost component in order to make profitable sales of its low-cost component. With unbundling, there is no such trade-off. This example illustrates a real-world effect, but it is just an example. With three or more firms, other examples can be constructed in which bundling leads to higher equilibrium prices, essentially because of the loss of mix-and-match benefits in terms of production costs (Farrell, Monroe, & Saloner, 1998).

### Vertical Unbundling Necessitates Standards That Stifle Innovation.

The need to set rigid interfaces to allow different firms' services to work together may have the effect of limiting innovation because new technologies

---

[25]Hence, the acquisition of market power through investment and hard work is not, in itself, illegal. (See, e.g., U.S. Department of Justice and the Federal Trade Commission, *Antitrust Guidelines for the Licensing of Intellectual Property,* April 6, 1995, § 2.2.) Moreover, intellectual property policy grants innovators and creators a degree of market power as an incentive.

[26]Algebraically, suppose Firm 1 has unit costs of producing the two components equal to $c_1$ and $d_1$. Suppose Firm 2 has unit costs $c_2$ and $d_2$, where $c_1 < c_2, d_1 > d_2$, and $c_1 + d_1 < c_2 + d_2$. With unbundling, Firm 1 makes all of the sales of the first component at a price of $c_2$ and Firm 2 makes all of the sales of the second component at a price of $d_1$. With bundling, Firm 1 makes all of the bundled sales at a price of $c_2 + d_2$, which is less than $c_2 + d_1$.

may not be readily compliant with the interfaces. Although the layered architecture of the Internet may lead to rigidities, these are likely to remain whether or not suppliers in the television industry are vertically integrated.

*Vertical Unbundling Leads to Consumer Confusion and a Loss of Supplier Accountability.* A somewhat different type of concern is that no one provider will have responsibility for the services ultimately delivered to consumers and/or consumers may be confused by the existence of different providers at different stages.

These concerns are misplaced. If vertical separation results in customer confusion or a lack of responsibility for customer satisfaction, then there will be market incentives for organizations to offer end users one-stop shopping even if the providers are not integrated or engaging in vertical bundling. Companies offering the one-stop shopping would take responsibility for end-to-end quality and for customer care. These companies would simultaneously enter into agreements with providers at various stages in the value chain specifying the responsibilities of each. As long as the interstage contracts did not call for exclusive dealing, the competitive benefits of vertical separation would be maintained even while offering one-stop shopping. Moreover, consumers may prefer vertical ownership separation and unbundling as means of ensuring objectivity in providing navigation services or recommending mix-and-match decisions to combine offerings at various stages.

The previous analysis is only a starting point. Vertical integration and bundling have a complex set of potential costs and benefits from both the private and social perspectives. Preliminary analysis, however, suggests that there are not strong arguments that extensive vertical integration and bundling are necessary or desirable to create investment incentives and facilitate coordination across stages of the television value chain.

## WINNERS, LOSERS, AND SURVIVORS

Who will benefit from the changes discussed earlier? And who will lose? The following discussion is organized around existing entities—rather than on specific stages in the value chain—both out of prurient interest and because existing entities are relevant decision-making units for both business and public policy analysis.

### Viewers

Ignoring the broad societal degradation that increased use of electronic media may bring about, the majority of viewers will likely gain significantly from the development of television over the Internet. Technological developments are making it possible to offer viewers a wider range of programming as well as programming with new features. These developments should also increase competition, which will ensure that the benefits of technological progress largely accrue to consumers, rather than suppliers. That is, increased competition for viewers' attention and money will almost certainly result in viewers' facing lower quality-adjusted prices. The fall in quality-adjusted prices will come about through a combination of lower prices and increased qualities. Increased quality, in turn, will be attained by a combination of programming with increased production values and content increasingly targeted to specific viewer interests.

Of course, not all viewers will gain. Consumers with a high tolerance for commercials who enjoy mass market programming may find that the new equilibrium is worse for them. One witnessed similar developments with the introduction of cable television. There are at least some instances in which sporting events now on cable television would have been on nonsubscription television if cable television had not existed. Viewers who receive strong over-the-air signals and care only about these events are made worse off by cable television.

### Advertisers

Like viewers, advertisers will enjoy the benefits created by technological progress because competition will drive suppliers to pass these benefits through to their customers. Targeted advertising and the ability to reach tightly controlled demographics will offer advertisers better services. In the other direction, advertisers will face threats from increased processing power in the hands of viewers. Audience fragmentation may not be a threat because, whereas audiences will continue to fragment, it is reasonable to predict that technology will create synthetic mass media.

### Local Broadcasters

The principal effect on local broadcasters of television over the Internet will be to devalue their key competitive asset—spectrum licenses—by creating substitute distribution channels. The television viewing shares of broadcast television have fallen steadily for the last two decades, whereas the viewing shares of cable and satellite services have risen.[27] Television over the Internet will continue this trend.

Today, local broadcasters do more than distribute content packaged by others. Local broadcasters create content, notably local news programming. This fact raises the possibility that local broadcasters could continue

---

[27]Among cable households, more than one half of their television viewing is now of cable networks and pay services, rather than programming that originated on a broadcast channel. (Paul Kagan Associates, *Cable TV Advertising*, February 28, 1999, and June 21, 1999.)

to dominate this role even after television migrates to the Internet. There are, however, at least three reasons to suspect that local broadcasters will not maintain a significant competitive advantage in news programming. First, the efficient geographic scope of newsgathering organizations may be national or international. At present, many local broadcasters rely heavily on network news organizations for significant programming. Second, reputation or brand is an important asset in the market for news programming. In many instances, this asset appears to belong to the national networks with which local broadcasters often affiliate, rather than to the broadcasters themselves. For example, the ownership of the NBC affiliate in San Francisco was transferred to Young Broadcasting in 2000. At the time, few viewers likely knew about this change of ownership, much less had an opinion of the new owner. Instead, viewers probably relied on the fact that the station still was an NBC affiliate to form their judgments about the likely veracity of the reporting.[28] Third, the opening of the distribution bottleneck will allow other firms with reputations and skills to enter the market. Newspapers, for example, are natural competitors to broadcasters in the provision of multimedia news sites and programming.

Local broadcasters will derive benefits from their current role in local news and public affairs coverage. Because of this role, local broadcasters are unquestionably one of the most powerful lobbying groups before Congress. Local broadcasters have repeatedly used this power to obtain regulatory protection from competition, whether from cable television or satellite. Because of the Internet's benefits are so far-reaching and the use of regulation to limit its range of applications is so difficult, competition from the Internet will very likely be impossible to thwart through the political process.

Faced with the impossibility of stopping television over the Internet, it is probable that over-the-air broadcast stations will be allowed to keep their spectrum but use it to provide other services. In some cases, this use will result in the first type of convergence identified in the introduction: the existing broadcasting infrastructure will become a carrier of e-mail, web traffic, data, and other "Internet" services. In other cases, the policy change may simply increase broadcasters' flexibility and allow them to offer traditional mobile voice or even subscription television using the spectrum to which they have usage rights. Although it might be more efficient to let broadcasters sell their spectrum rights to firms more capable of offering nontelevision services, doing so would make it harder to justify the spec-

---

[28]Author's update: In January 2002, Young Broadcasting's San Francisco station ceased being an NBC affiliate. In November 2002, Young Broadcasting reported that its station was the number one news station in the San Francisco market. Available at http://www.youngbroadcasting.com/ireye/ir_site.zhtml? ticker=YBTVA&script=2100. These facts suggest that either reputations attach to on-air personnel (who remained largely the same after termination of the network affiliation) or reputations are unimportant.

trum rights giveaway. A cynical forecast is that local broadcasters will use the spectrum themselves "to serve the public," and they will do so by taking on partners that have the skill sets needed to offer these services.

## Cable Companies

Technological trends will work both for and against cable systems operators. With their fiber-coax networks, cable systems are leading candidates to evolve into distributors of television over the Internet. Thus, cable systems operators will benefit from having increased systems capabilities and thus being able to offer more attractive services to their customers. However, to the extent that alternative forms of video Internet access develop (e.g., if telephone companies' DSL becomes television capable), cable companies will face increased competition for their traditional services, as well as any new ones.

## Broadcast Television Networks

Today, broadcast television networks are involved in content creation, packaging, and distribution. Thus, these companies will almost certainly survive the transition to Internet television (although some may be acquired as part of ongoing industry consolidation). These companies will create content and package it for a variety of distribution formats. These developments will continue trends already under way. In response to the rise of cable and satellite multichannel video, over-the-air broadcast networks and their parent companies developed or acquired cable properties. For example, ABC's parent owns ESPN; CBS owns the Country Music Channel; and Fox has Fox News, FX, and the Fox Movie Channel. In response to the rise of the World Wide Web, over-the-air broadcast networks and their parent companies developed dozens of Web sites. For example, ABC's parent owns ESPN.com, one of the top sites on the Internet, and CBS owns MarketWatch.com and MedWatch.com.

## Independent Content Producers

Content producers will no longer be squeezed through a distribution bottleneck. Consequently, there will be an increase in demand for programming that has intense appeal to narrow audiences. The history of videocassettes and the World Wide Web suggests that pornography will very likely see an increase in demand, for example. Whether content producers will earn large rents is less likely, however, because of the monopolistically competitive conditions that are likely to prevail for niche programming given product differentiation and the large number of potential producers.

Producers of programs with mass appeal face more mixed effects. Anecdotal evidence suggests that the broadcast rights to major sporting events, awards shows, and other "event" programming have become dramatically more expensive as broadcasters have competed for programming that can attract mass audiences in a multichannel world. However, synthetic mass audiences may devalue these skills, and mass market content creators will face increased competition for viewers' attention from niche programming.

## CONCLUSION

This chapter offers several predictions about the future of television over the Internet. In closing, I want to be clear that I am speculating about the *distant* future.[29] It is safe to say that the widespread deployment of Internet television will take longer than many people think.[30] In 2010, the majority of viewers will be watching television that is largely as it is known today and is received either over-the-air or on cable systems constructed primarily to broadcast video. Although the penetration of the Internet has been impressive, it is still far below that of television and shows no sign of approaching it anytime soon.[31] Moreover, the penetration of broadband last-mile access (in the form of cable modems and DSL) is much lower still, and does not offer broadcast quality video in any event.[32]

What will happen over the next 10 years? Optional interactivity will increasingly be offered as a supplement or enhancement for broadcast and

---

[29]Noam (1995) made many of the same predictions for what was then "the future" that I and others are making today. In part, this is a testament to his foresight, and, in part, it is a reflection of the fact that Internet television has developed very slowly over the last 7 years.

[30]It is amusing to read with the benefit of hindsight the various predictions about interactive and Internet television made during 1994 and 1995 by industry members and analysts.

[31]Television's penetration of U.S. households stands at more than 95%. In August 2000, 41.5% of U.S. households had some form of Internet access. It is notable, however, that penetration rates are considerably higher for high-income households, who presumably are the most commercially attractive viewers and subscribers for television over the Internet. (Economics and Statistics Administration, National Telecommunications and Information Administration, *Falling Through the Net: Toward Digital Inclusion—A Report on Americans' Access to Technology Tools*, October 2000, at 1 and 8.)

[32]At the end of 1999, there were only 1.8 million residential subscribers to telecommunications services capable of delivering transmission speeds of 200 kilobits per second or more in at least one direction. (Federal Communications Commission, *In the Matter of Inquiry Concerning the Deployment of Advanced Telecommunications Capability to All Americans in a Reasonable and Timely Fashion, and Possible Steps to Accelerate Such Deployment Pursuant to Section 706 of the Telecommunications Act of 1996*, "Second Report," CC Docket No. 98-146, released August 21, 2000, at ¶ 8.) In contrast, there are approximately 100 million television households.

cable programming. Consumers will enjoy control of increasing process-ing and storage power, providing the industry a taste of asynchronous viewing and various threats to advertising business models. Satellite and cable capacity will continue to increase, offering ever greater program-ming variety. In other words, television will undergo an evolutionary pro-cess, "Internet time" notwithstanding.

## REFERENCES

Computer Science and Telecommunications Board, National Research Council (2001). *The Internet's coming of age.* Washington, D.C.: National Academy Press.

Cournot, A. A. (1838). *Researches into the mathematical principles of the theory of wealth,* English translation of French original. New York: Kelly.

de Moragas Spa, M., Garitaonandia C., & Lopez, B. (1999). Regional and lo-cal television in the digital era: Reasons for optimism, In M. de Moragas Spa, C. Garitaonandia, & B. Lopez (Eds.),*Television on your doorstep: Decentralisation experiences in the European union.* Luton: University of Luton Press.

de Vos, L. (2000). "Searching for the Holy Grail: Images of Interactive Tele-vision," dissertation, University of Utrecht. Retrieved September 7, 2000, from http://www. globalxs.nl/home/l/ldevos/itvresearch/

Farrell, J., & Katz, M. L. (2000, December). Innovation, rent extraction, and integration in systems markets. *Journal of Industrial Economics, 48:* 413–432.

Farrell, J., Monroe, H., & Saloner, G. (1998, Summer). The vertical structure of industry: Systems competition versus component competition.*Journal of Economics & Management Strategy, 7*: 143–182.

Faulhaber, G. R, & Hogendorn, C. (2000, September). The market structure of broadband telecommunications. *Journal of Industrial Economics, 48*: 305–329.

Katz, M. L. (1989) Vertical contractual relationships. In R. Schmalensee & R. D. Willig (Eds.), *The handbook of industrial organization. Amster-dam: North Holland Publishing.*

Griliches, Z. (1992). The search for R&D spillovers. *Scandinavian Journal of Economics, 94*(Supplement): 29–47.

Grossman, S., & Hart, O. (1986, August). The costs and benefits of ownership: A theory of vertical and lateral integration. *Journal of Politi-cal Economy, 94*: 691–719.

Jones, C., & Williams, J. (1998). Measuring the social return to R&D. *Quar-terly Journal of Economics, 113*(4): 1119–1135.

Nalebuff, B. (2000). Competing against bundles. In P. Hammond & G. Myles (Eds.), *Incentives, organization, and public economics: Papers in hon-our of Sir James Mirrlees. Oxford: Oxford University Press.*

News Corporation *Magazines and inserts.* Retrieved December 17, 2002, from, http://www.newscorp.com/operations/magazines.html

Noam, E. M. (1995). *Toward a third revolution of television.* Retrieved October 31, 2000 from http://www.vii.org/papers/citinom3.htm

Owen, B. M., & Wildman, S. (1992). *Video economics,* Cambridge: Harvard University Press.

Perry, M. K. (1989). Vertical integration: Determinants and effects. In R. Schmalensee & R .D. Willig (Eds.), *The handbook of industrial organization.* Amsterdam: North Holland Publishing.

# 5

# Business Models and Program Content

David Waterman
*Indiana University*

Beginning in the mid-1990s, the Internet unleashed an extraordinary amount of experimentation with the delivery of broadband entertainment content to consumers. Much of that content has been Internet-original, notably short films and serials, and interactive program forms. At the other end of the spectrum have been feature films and TV programs already appearing in theaters or on other media. The suppliers of this content have experimented with just as wide a range of business models: advertising, sponsorship, bundling with other products, promotion of other products, instant online purchase of merchandise, pay-per-view or "rental," sale of content (by consumer downloading), and hybrid forms.

Quite a few of the innovators have failed along the way, and for those who have survived the dot.com bust or who have entered the market since, ambitions have been tempered. Few observers, however, would discount the long-term possibilities for delivering broadband entertainment over the Internet. From that perspective, several big questions immediately come to mind. What kinds of business models will predominate when Internet television eventually develops? How will file sharing technologies affect these business models? And what types of content will these business models support: a wealth of new niches, or just more of the same?

Of course, no one can answer these questions with any certainty. Indeed, arriving at the answers is the market function of all the experimentation in the first place. The objective here is much more modest. By applying some economic principles, and considering the historical experience with established broadband media, this chapter provides a framework for thinking about the answers to these questions.

## SCOPE

To begin, some subject matter boundaries. First, the focus is entirely on actual Internet protocol (IP) delivery of television programming. For example, "enhanced" television, using the computer (or a set-top box) to interactively play along with a standard television exhibition of a sports event or game show, does not count. Second, the discussion concentrates mainly on dramatic entertainment forms for consumers. There are many business-to-business (B2B) broadband applications (e.g., video-conferencing), business-to-consumer electronic commerce that is unaffiliated with entertainment programming, interactive computer games or gambling, pornography, and news programming—all of which have evident economic potential on the Internet, but are also outside the scope. Third, the concern is for the long-run future. Internet bandwidth capacity, as well as payment mechanisms and general usability of computers, will have to develop far more for Internet TV to reach its potential. There is an assumption that those developments will eventually happen, but no predictions on when are made. Another assumption is that in the long term, the computer and the television set will converge. That reflects the faith that if Internet TV entertainment technology develops in a potentially profitable way, then the computer will find its way into living rooms. Finally, the focus is primarily on developments in the United States.

The next section begins with a brief review of the academic literature preceding this study. Then an overview of Internet TV entertainment experiments is provided. After that, five basic economic characteristics of the Internet that are relevant to development of Internet TV business models and content are set out. Implications for future development of business models and broadband entertainment content are then given.

## LITERATURE

Scholarly discussion of the economics of Internet television has been sparse, but several works provided a useful foundation for the present study.

Owen (1999) took a basically pessimistic view of the future of Internet television, arguing not only that adequate bandwidth appears to be far into the future, but that the architecture of the web is not well suited for broadcasting video. Tristam (2001) also discussed a number of current and future economic and technological limitations of broadband video. The Communications literature has also been concerned with Internet TV. Kiernan and Levy (1999) studied the content of broadcast related Web sites, for example, and a series of articles on the future of the Internet (notably Shaner, 1998) have conceptualized the nature of web content on a broader level. Of more direct relevance to the present analysis, Picard (2000) offered a study of the historical development of business models for online content providers more generally, and suggested lessons in these experiences for the future. Konert (2000) analyzed a variety of financial and revenue generation models for Internet broadcasting from a European perspective, especially with regard to implications for European public broadcasting. Shapiro and Varian (1999) discussed economic characteristics of digital and networking technologies, including the Internet, from the standpoint of advising business people how to take advantage of these technologies and design better business models. Bakos and Brynjolfsson (1999) studied the economics of product bundling strategies on data networks. A recent National Research Council report, *The Digital Dilemma* (2000), discussed in detail the economics and technology of the Internet from the standpoint of copyright and other government policy questions. A number of law review and other policy-oriented papers, such as Samuelson (1999), Schlachter (1997), Einhorn (2000), and Jackson (2001), also discuss economic and technological characteristics of Internet entertainment delivery from those perspectives.

## INTERNET TV EXPERIMENTATION

A number of commercial Web sites offering Internet TV entertainment in the past few years—some of them now defunct—illustrate the wide variety of business models and content that entrepreneurs have experimented with in the United States.

On the content side, film shorts and serials, or "webisodes" (also mostly very short), have been pioneered by sites such as ifilm.com, atom films.com, icebox.com, and entertaindom.com. Many of these programs have been originally produced for the Internet. The great majority of serials seem to be animation, though there have also been talk shows (e.g., Cyberlove on thesync.com). Other sites, such as sightsound.com, ifilm.com, and later cinemanow.com (along with file sharing sites such as kazaa.com and grokster.com) have offered access to feature films. Most of these have been theatrical films, although one Internet-original feature distributed by sightsound.com in 2000, *The Quantum Project,* received a lot of publicity. The sightsound.com site later began offering recent full-length features from the Miramax studio (owned by Disney), and cinemanow.com followed with other Hollywood fare. A consortium of five major studios then began offering newer features via movielink.com, and other studios are following. Among other innovations, itsyourmovie.com experimented with an original serial program in which viewers could vote on the future plot direction. Sony's site, screenblast.com, allowed users to interactively create their own mini-episodes of popular TV programs.

Many broadcast TV stations have been streaming their programming from Web sites for several years, and broadcast.com has retransmitted television stations worldwide. Many other sites, such as CBS.com and NBCi.com, have offered a wide variety of short video news clips, previous episodes of entertainment series programs, and of particular interest, outtakes of other original clips that supplement regular TV series (e.g., CBS's *Survivor*).

The range of business models employed by these Internet broadband sites has been great. Banner advertising and its increasingly proactive forms, other links to retail outlets, and on-site merchandising have become common components. Brief "pre-roll" commercials have become routine preamble to entertainment content. Users can often click on these commercials and get more product information. Some sites, such as BMWfilms.com and skyy.com, have offered high production quality short films that overtly promote their sponsors' products (namely, BMWs and Skyy Vodka) by integrating them into the stories. Instant online purchase of products modeled within an entertainment program was apparently first experimented with by Microsoft. Internet TV promos of programming available on other media, especially broadcast or cable programs and movies, have been very common fare on broadband sites, and seem to drive the economics of broadcast and other traditional media-based sites.

Led by sightsound.com, broadband entertainment sites have increasingly moved toward direct payment models. Growing numbers of older and especially recent feature films have been available for a 1- or 2-day pay-per-view license, usually for $2 to $4, or purchase typically for $5 to $15.

## FIVE ECONOMIC CHARACTERISTICS OF THE INTERNET AFFECTING INTERNET TV BUSINESS MODELS AND CONTENT

The idea that the Internet is a revolutionary communications medium has become common currency in discourse about the media. This label is certainly justified in some contexts. From this chapter's perspective, the Internet is best viewed in comparison with established broadband media in terms of the economic *improvements* it can make to cost and efficiency features of those media.

The Internet's economic improvements on established media can be divided into five categories: lower delivery costs and reduced capacity constraints, more efficient interactivity, more efficient advertising and sponsorship, more efficient direct pricing and bundling, and lower costs of copying and sharing.

## Lower Delivery Costs and Reduced Capacity Constraints

Media transmission system costs consist of several components: a capital infrastructure for transmission, home premises equipment, and variable costs of delivering the information. Parts of these infrastructures and home equipment have multiple uses, and costs often depend critically on usage rates. Cost comparisons among media are thus difficult. Some comparisons show that Internet transmission of television signals is currently far more expensive than cable and some other media (see Noam, 2000; Noll, 1997); Owen (1999) and Tristam (2001) argued that for mass distribution, broadband streaming over the Internet will not for the foreseeable future be as efficient as broadcast television, although IP costs continue to fall. In other respects, the Internet has major cost advantages. Internet TV is more-or-less free of geographic constraints, allowing essentially instantaneous worldwide transmission. A component of delivery costs is the ability of consumers to simply download content rather than to copy in real time off of a cable channel, for example, or take a trip to the video store to buy a product that has been manufactured, packaged, shipped, and maintained in an inventory. From the latter perspectives at least, Internet transmission of video is quickly becoming more cost efficient than existing media.

Because of its architecture, capacity or "carriage" constraints become very minor on the Internet. In the 1940s, increasing the number of available movies in a town meant building a whole new theater. Broadcast TV stations reduced these capacity costs, especially in larger markets. Cable TV and DBS have further reduced capacity constraints, and these costs continue to fall with digital compression technologies. Video stores have essentially the highest "capacity" of any established broadband media. All of these media, however, have significant carriage costs. Another channel on a cable system requires a major investment, even with digital compression technologies. Another video or DVD at retail stores requires total demand for a few thousand copies to make duplication and physical distribution worthwhile. The stores that carry each title must cover inventory costs for as long as consumers wish to rent or buy it. On the Internet, a variety of Web sites can offer a virtually unlimited number of products, and consumers can readily switch within and among different sites.

The implication of these cost and capacity advances is lower prices and, especially, greater product variety. That variety provides one ingredient for virtually "true" video-on-demand systems. Also, thinner and more marginal markets can now be served.

The latter potential is shown by the abundance of Internet-original short films and serials already available. One factor is probably just their suitability to a medium in which more lengthy viewing or downloading experiences are now too tedious. Many have earned critical praise, however, and their often racier content is generally differentiated from other broadband content. Nevertheless, another economic reality underlies their prevalence on Internet TV: consumer demand for short subjects has in the past usually been too marginal for all but a few to even be made available in specialty video stores or on the most narrow appeal cable television channels.

Another example of relatively marginal content on the web is movie or program outtakes: footage about the making of a program, interviews with the creative people, and so on. Currently, such material is included as extras on some DVDs, but as some Web sites are already demonstrating, the Internet expands these possibilities almost without limit. The situation is similar for supplementary material about advertised products. Such ancillary video materials are important building blocks for both advertising and direct pricing business models.

## More Efficient Interactivity

If the Internet has a forte among its many marvels, it is surely two-way interactivity. Interactivity has been physically possible since cable systems offered it in early years, notably on the QUBE system in Columbus, Ohio, in the 1970s. Also, a hybrid form of interactivity is now available with the integration of computers and standard TV transmissions to create enhanced TV. Viewers with a computer in the same room (or a set-top box) can simultaneously play along with game shows or sports events. Viewers can also buy products shown on standard television commercials more and more easily with the right home equipment. In some systems, viewers can now choose between several simultaneous feeds of standard broadcast content (e.g., different camera angles covering a sports event) to control the pictures that they actually view. Personal video recorders (PVRs) permit asynchronous control of programming starts and stops.

Cable, DBS, and other multichannel systems are rapidly developing interactive technology as well. But the development of Internet TV should permit most of these activities to be conducted more efficiently. Viewers can instantly and easily control a much wider variety of programming content via their responses. Home shoppers can simply click on a product shown in the middle of a televised movie to instantly buy it, or to get more information about it. They can do the same with an in-show TV commercial. More efficient interactivity offers another ingredient of true video-on-demand systems as well: a convenient process of ordering movies or other programs for on-screen viewing or for download.

With Internet technology, viewers can also neatly manipulate the sequencing of video images. Many question whether consumers in any significant numbers will (apart from the case of pornography) ever want to fiddle with the narrative form of entertainment programs. Still, a great amount of innovation is being invested in systems that will allow people to have that option.

## More Efficient Advertising and Sponsorship

A perennial limitation of television advertising has always been waste circulation because of muddy demographic, product interest, or other segmentation. Cable and other multichannel systems have reduced this problem by making room for more sharply targeted programs. Internet television permits the chance to further advance this quest in two ways. First, the virtual removal of capacity constraints should allow still sharper segmentation in the same way that multichannel systems have improved the broadcast model. Second, the ability of advertisers to track the buying or Internet usage patterns of individual consumers permits different ads to be inserted within (or displayed alongside) the same program, depending on the viewer's revealed interests or estimated willingness to buy a particular product.

The click-through interactive system of Internet advertising is much like the per-inquiry (PI) ads often seen on cable TV networks, in which the network is paid not for exposures, but earns a percentage of each purchase made via a phone number displayed on screen. The Internet system is a more efficient PI system. Even without click-through purchasing opportunities, the ease of obtaining more information about products with a mouse-click is a significant advance in product information dissemination. Finally, the Internet offers the opportunity for full sponsorship of a Web site, or of an area within a site, that attracts consumers with entertainment programming. This sponsoring system might meld the branding of a dramatic format program and its characters with a consumer product in a better way than television program sponsorship, first developed in the late 1940s, has been able to do in the past.

Although Internet technology thus promises more efficient advertising-based business models to support broadband programming, history suggests formidable practical limits. First, although multichannel cable television has brought forth billions in total advertising, including many new advertisers, the "magazine model" of higher rates for sharper segmentation has not materialized. A few networks, such as MTV, have segmented very successfully, but cable network cost-per-thousand ad rates are on average still well below those of the major broadcast networks, apparently due mostly to the limited national audience reach of networks that rely on multichannel system delivery (Media Dynamics, 1999; Waterman & Yan, 1999). Second, whereas in-show commercials can be carried on Internet TV programs, the click-through potential on the Internet does not seem to offer a great advantage over what virtually ubiquitous broadcast television stations already do with in-show commercials, especially given the relative importance of product image advertising. Most products are not subject to impulse purchase or PI models of advertising. Third, as seen with VCRs and now with PVRs, consumer control digital technologies like the Internet generally increase the ease with which viewers can zap ads or otherwise avoid them.

Overall, the success of advertising as support for Internet broadband entertainment seems to rely heavily on consumer initiative to investigate or make online purchases of advertised products—a plausible model, but one with a spotty historical record on other media. Furthermore, to the extent that Internet TV evolves into a "store and replay" rather than "live" transmission medium, as some believe will happen, advertising's potential will also be limited (see Odlyzko, 2000).

Analysts' initial expectations for the potential of Internet advertising models have greatly diminished (for the aforementioned and perhaps other reasons). Nevertheless, innovation is active, and at least some of the potential improvements to advertising efficiency on the Internet should materialize. For at least some products, the result should be more cost-effective advertising and product promotion, and thus an increase in the effectiveness with which advertising and sponsorship can support broadband entertainment content. More sharply focused programs should accompany these developments.

## More Efficient Direct Pricing and Product Bundling

More efficient direct pricing means lower costs in making transactions, but especially the ability to more effectively price discriminate—that is, to extract the maximum amount that each consumer is willing to pay for a product. In several respects, Internet technology promotes these efficiencies.

First, direct payment-supported video-on-demand systems are likely to evolve to be at least as cheap and easy to manage by Web sites as they ever will be on cable or DBS. Micropayments, which allow very small amounts (perhaps only a few cents) to be automatically charged to a user via a credit card or similar means, are a prospective component of true video-on-demand systems, although they have recently encountered development problems.

Web sites can also engage in so-called dynamic pricing, by which consumers are charged different prices according to their perceived willingness to pay, based on prior purchasing habits on the web, Web site visiting habits, or other information. Basically, dynamic pricing permits more efficient price discrimination through better identification of high versus low value customers.

An important component of effective direct payment systems is efficient bundling of products, such as a package of three movies together, monthly subscriptions, or the sale of movies along with talent interviews, outtakes, and so on. A large literature in economics has explored many ways that such packaging can extract consumer surplus via price discrimination (e.g., Adams & Yellin, 1976; Varian, 1989). Of course, video stores and cable- or satellite-based systems also offer bundles. But on the Internet, tailor-made packages can be offered to different consumers depending on buyer profile data, and interactivity allows choice among more complex menus or package variations than other media can efficiently offer.[2]

A variety of other price discrimination devices, such as reduction of prices over time for movies as they become older, or lower prices for repeat viewings, are also efficiently managed on the Internet. A plausible method of Internet TV price discrimination may involve consumer segmentation based on demands for different qualities of transmission. Consumers with higher speed connections, for example, are likely to have higher valuations for high technical quality.

As with advertising-based business models, these potential improvements in direct pricing have practical limits. There is a long history of apparent consumer resistance to paying at every turn (e.g., the failure of DiVX, the digital videodisc system promoted by Circuit City in the United States that allowed consumers to pay according to the number of times a program was watched). More generally, pay-per-view systems have not done very well on cable or satellite systems, although it is unclear how much the lack of consumer control over starts and stops, the limited selections, or other factors are responsible. Dynamic pricing may also have an uncertain legal future. Also, even though the Internet theoretically allows practically any kind of segmentation to take place, it may also prove difficult to price discriminate geographically with an inherently nationally and internationally distributed medium. Geographic discrimination is a natural process for video stores and cable systems.

Undoubtedly, some of the theoretical advantages of direct pricing on Internet TV will never happen. The Internet offers such potential in this area though, that at least some of its advantages, in terms of lower transactions costs and more efficient market segmentation, seem bound to become established. The result should be more viable VOD systems and greater revenue support for products with relatively high consumer demands.

### Lower Costs of Copying and Sharing

Attracting more recent attention than any other attribute of the Internet is the remarkable ease with which content, including movies or other videos, can be duplicated and transferred from one consumer to another. The popularity of Napster and gnutella-like file sharing systems have been a testament to these efficiencies. The limited use with video content on

---

[2]Bakos and Brynjolfsson (1999) studied the economics of offering menus of very large bundles on the Internet.

these systems thus far is no doubt due largely to bandwidth constraints. Of course, copying and sharing of movies and other videos has been widely practiced since VCRs arrived along with copy-prone pay-TV movies and prerecorded cassettes that can be copied back-to-back. The consumer's task of copying and sharing simply becomes far less time consuming and awkward with the use of a computer.

As everyone has recognized, computer network technologies for copying and sharing pose a serious threat to copyright holders because paying customers can practically evaporate from the market. Even a single casual file transfer can have devastating cumulative effects as it is retransmitted from user to user virtually without cost.

Attracting increasing attention are the new opportunities for copyright owners created by efficient duplication and file sharing via the Internet through digital rights management (DRM). Already mentioned are the negligible costs of a consumer download—essentially equivalent to copying—compared to purchasing a DVD or videocassette, or of making a real-time copy off of standard television. With existing pay-per-view or home video systems, consumers who want to share a copied movie with someone else also have to physically deliver it to the recipient. Peer-to-peer computer transfer essentially eliminates that cost. Fundamentally, the lower consumer costs of copying and peer-to-peer transfer via the Internet create market value. If distributors can manage to appropriate some or all of that created market value, their revenues and profits will rise (Besen, 1986). Consider the "old" system in which consumers have made real-time back-to-back copies off of prerecorded videos or off pay-per-view channels to share with others. The copyright owner may be able to appropriate some fraction of the value of that physically shared copy to the recipient, but it is almost certainly lost revenue for the most part.[3] If it is assumed for the moment that distributors are able to maintain strong copyright protection governing broadband Internet transmissions, then they may be able to appropriate all, or at least a larger part, of the value of an electronically shared copy. For example, an automatic electronic payment to the distributor could be activated by a peer-to-peer file transfer (e.g., via a gnutella or Napster-like system) of a copyrighted movie.[4] Alternatively, such peer-to-peer file transfers could be forbidden by copyright owners, and all users simply induced to purchase directly from the owner. In these eventualities, the incentive for consum-

---

[3]Besen made the unrealistic assumption in his model that the distributor can appropriate all of this value. In fact, the most that the distributor can ordinarily appropriate is the value of the product to the buyer plus the value that buyer realizes from making and distributing copies. The latter component is likely to be less than the value of the copies to those who receive them. See Katz (1989) for a useful analysis of home copying issues from an economic perspective.

[4]One indication of this potential is that the movie site sightsound.com has experimented with using gnutella.com to deliver encrypted movie files to users, who in turn paid sightsound.com a fee for the key (Snell, 2001).

ers to engage in the cumbersome process of physical copying and sharing will also be reduced to the extent that prices for authorized electronic download or peer-to-peer file sharing are low enough to render the physical process a less desirable alternative.

Internet technology thus increases the distributors' potential revenues from movie or other product distribution. Possibly, these revenues can be enhanced by improved price discrimination as well. Those who take advantage of file sharing probably tend to have lower price demands, and thus may drop out of the market at the distributor's price for the "original" movie. If distributors can devise a method for charging lower prices for movies transferred from peer-to-peer file sharing sites, or from other sites that involve greater consumer search costs, than for direct downloads from the distributors' sites, they could also increase revenues.[5]

Of course, it is unrealistic to believe that e-mail or other unpaid peer-to-peer transfer of movies and other broadband entertainment could ever be eliminated, even if these practices were made illegal. Also, Internet distribution of movie data stripped from DVDs remains a dramatic threat. However, watermarking and other copyright protection technologies for authorized Internet distribution are rapidly developing, and the recent entry of Hollywood studios into Internet distribution of their movies suggests improved technologies of protection. If copyright interests continue to get favorable court interpretations of the 1999 Digital Millennium Copyright Act's prohibition on attempts to defeat encryption, and new legislation is enacted to account for newly developing problems, then copyright owners should be able to keep losses to a minimum.[6]

The historical experience with back-to-back video copying by consumers encourages that speculation. Surveys indicate that consumer sharing of back-to-back video copies accounts for only about 1% of legitimate market transactions, and the overwhelming proportion of consumers believe

---

[5]Before Napster's demise, its negotiations with Bertlesmann for Napster to price their music services to consumers were headed for just such a price discrimination system. According to press reports, a $2.95 to $4.95 monthly subscription price was reportedly being discussed for a fixed number of music file transfers on Napster. For $5.95 to $9.95, unlimited transfers could be made. Additional charges would be made for the right to record the music onto blank CDs. The technical quality of all these paid subscriber transfers or recordings, however, would only be "near-CD" quality. Thus, higher value consumers would be induced to pay progressively more to use the service, but restrictions on transmission quality would still serve to segment the higher value CD and lower value file-transfer market segments. (Clark, 2001). Of course, the well-publicized resistance of other record companies to Bertlesmann's proposals suggests that such a direct pricing system was too clumsy or impractical in the current environment.

[6]See Jackson (2001) for a concise discussion of the DMCA and its application to the Napster case.

that back-to-back copying is illegal, suggesting that relatively small minorities would try to defeat encrypted programming even if they could, or would make illegal peer-to-peer transfers.[7] Those who do are likely to be low value consumers who would be disinclined to pay for the programs at the prevailing retail prices in any case.

In summary, Internet technology offers many ways by which program distributors can not only reduce costs of delivery and improve desirability of the programming packages they offer to consumers, but also improve the advertising, direct pricing, and other components of their business models. Web entrepreneurs are already combining components of business models in imaginative ways. Undoubtedly, some of these potential improvements will not work out. The law, the advance of technology, and uncertain demand could all inhibit them. But the potential of the Internet seems so great in these respects that it is hard to imagine that Internet TV will not—at least eventually—lead to some marked improvements in television distribution and the business models that support it.

## IMPLICATIONS FOR BUSINESS MODEL AND CONTENT DEVELOPMENT

### Advertiser Versus Direct Pricing Support

Internet television has been disproportionately reliant on advertising or e-commerce related business models although no one seems to claim profits to date with any model. The shift toward direct-payment models now underway is likely to continue for two reasons. One is that the intense competition among web distributors to establish themselves in the market during the Internet's growth stage surely inhibited many firms from charging directly. The second reason to expect more direct pricing is that higher bandwidth capacity will mean that products of greater consumer value (viz., feature-length movies and sporting events) can be attractively presented. Other than pornography, consumers have never been willing to pay directly for much audio/visual entertainment besides movies and some sports. Historically, advertising has mostly been used to support content watched by low value viewers who are unwilling to pay enough to outdo the few cents per viewer that advertisers will pay for an exposure. Although more efficient advertising and e-commerce related systems are likely to increase the value of Internet exposures to advertisers, it seems unlikely that these improvements will overcome the basic economic forces guiding high value viewers toward direct payment systems.

---

[7]The video copying percentage is derived from Office of Technology Assessment (1989) and Macrovision, Inc. (1996). The Macrovision study also reported that over 95% of survey respondents said they believed back-to-back copying of prerecorded videocassettes is illegal.

## The Internet as a Component of Multimedia and International Syndication Models

As broadband media have proliferated in the past two or three decades, individual programs are more frequently distributed on several different media over a period of time. As countries throughout the world have privatized their media and relaxed trade barriers since the mid-1980s, international markets, especially for U.S. entertainment products, have also expanded.

As Internet TV develops, there will be tremendous economic pressures for the providers of its content to employ similar multimedia syndication models, as well as to supply products that have international appeal. In brief, higher revenues can be generated both because total potential audiences can be reached and because audiences can be more efficiently segmented. The result is greater potential revenues, which will support higher production investments, and in turn attract larger audiences. Another factor favoring multimedia syndication is marketing. A high expenditure ad campaign supporting a product release on one medium serves to increase demand in all the products' potential syndication markets, and thus realize economies of scale in the same way that high production investments can be spread over large potential audiences.

The best illustration of the compelling economic logic of multimedia syndication models is the current system of theatrical feature film distribution.

*Movie Distribution and Internet TV.*   Everyone is generally familiar with the process by which movies are released over time in sequence to theaters, then to hotels and airlines, to videocassettes and DVD, to pay-per-view television, to monthly subscription pay TV, and finally to television broadcasting or basic cable networks. It is widely recognized that this release sequence is basically a method of price discrimination.

The key requirement for any price discrimination is the ability to segment high value from low value consumers. The movie release sequence appears to involve two main segmentation devices. The first is time separation between release to different media. High value consumers having intense demand for a particular movie (or movies in general) are induced to pay higher prices for a first-run theatrical exhibition, while other viewers wait for video, pay TV, or later exhibitions. The second segmentation device in movie distribution is product quality. In general, a theater offers a higher quality exhibition than does a TV exhibition. Similarly, the ability of a VCR or DVD player to stop and start a movie, the absence of commercials on PPV, and so on are quality attributes that attract higher value consumers. The end result is that effective prices paid by different consumers in the release sequence tend to drop over time.

According to some, the business models of movie distributors will have to change and adapt with the Internet, but Internet TV actually fits naturally into this ready-made model. If effective unbundled direct pricing models evolve, and piracy of the Internet distributions themselves does not prove overwhelming, then Internet movies will probably be exhibited in a similar window to that currently occupied by PPV or by video rentals and sales. Internet advertising models will probably be less valuable to movie distributors, but to the extent they do prove efficient, movies can be released on the Internet with advertiser or other commercial support, presumably at a later point in their business life. Precisely where the Internet fits into the movie distributors' business models depends on uncertain technological, legal, and demand developments, and will evolve from experimentation.

Wherever the Internet eventually fits, it is unlikely to replace other movie media in the sequence. All of them, from video stores to pay cable systems, have different quality attributes or different demographic appeals that further the distributors' objectives of segmenting markets in order to charge different prices for essentially the same product. DVD or videocassette retailers, for example, offer services that may never be effectively duplicated by the web. The physical search and human interaction in shopping for videos may have inherent advantages, as do the joys of physically owning a professionally packaged DVD or a tape. A related advantage of retailers is gift marketing, for which a well-packaged physical object is highly valued.

Theatrical film distribution also demonstrates the advantage of multimedia marketing. Advertising and publicity campaigns sometimes costing as much or more than the distributor grosses from theatrical exhibition are launched to support a theatrical release. Much of the benefits from this campaign are reaped as the film travels to video, pay TV, and other media in the subsequent months and years.

*Syndication of Internet-Original and Other Products.* The market segmentation/price discrimination opportunities for multimedia syndication are not confined to movies. Many programs, including broadcast network programs, made-for-pay (monthly subscription) movies and series, direct-to-video features, and made-for-(basic) cable programs, all depend heavily on syndication to other media, in domestic and foreign markets, and they maximize their revenues by similar means of market segmentation.

Along similar lines, there is certain to be a wealth of "Internet-original" programming that is exhibited later, or even simultaneously, on other media. The business model of atomfilms.com, for example, is already heavily dependent on multimedia syndication. In addition to its Web site-based advertising, atomfilms.com has distributed collections of its best short films to cable networks, airlines, and other media, and compiles them onto DVDs and videocassettes for rental and sale as well (Long, 2000). It was estimated in late 2000 that atomfilms earned two thirds of its total revenues from these "offline" sources (Mathews, 2000).

### Empirical Comparisons

*Multimarket Syndication and Programming Budgets.* The resulting economic advantage of program syndication over time is simply that larger program budgets, and thus programming with higher production values, can be supported. These more expensive programs attract larger and higher paying audiences. The contrasts in program investments of various entertainment products in the United States are illustrative. Although there can be wide variance, the average major Hollywood studio theatrical feature was reported to cost about $48 million in 2001 (MPAA, 2002). Based on recent trade reports, HBO's made-for-pay feature films average something over $8 to $10 million, made-for cable and made for broadcast features cost $3.5 to $5 million, 1-hour network dramatic series average $1.5 to $2 million per episode, a basic cable drama averages $750,000 to $1.2 million per episode.[8] These programs depend heavily on aftermarket syndication to support their investments levels, and there is a general correspondence between these programs' budget levels and their viability in aftermarkets.

By contrast, a report in *Variety* estimated budgets for 3- to 5-minute webisodes at approximately $10,000 to $20,000 (Graser, 2000). Some Internet-original long feature projects have been reported to cost in the $80,000 to $100,000 range. *Quantum Project* cost approximately $3 million for about 36 minutes of entertainment, and its goal was clearly to gain publicity for sightsound.com (Chetwynd, 2000). Of course, these relatively low budgets partly reflect the currently low household penetration rates of broadband Internet capability. They emphasize the point, however, that although creativity and imagination can go a long way on a shoestring, and sometimes lead to extraordinary results, the most successful Internet television programs are likely to be those that can be successfully adapted and sold to other media.

*The Videocassette and Cable Experience.* The economic significance of syndication-based business models is illustrated by the experiences of prerecorded videocassettes/DVD and cable television in the United States.

Video content, as described by the time usage patterns for home video in Table 5.1, is dominated by feature films and children's programs. As a visit to any video store shows, a vast number of obscure, narrow appeal movies, how-to, and other programs are also available on video, but the

---

[8]These data compiled from *Variety*, March 6–March 12, 2000, p. 58: TNT taps DeBitetto Originals Prexy; Schneider et al (1999), The Green Behind the Screen, *Electronic Media*, August 2.

**TABLE 5.1**
*VIDEOCASSETTE/DVD CONTENT DATA*

| Time of Use by content (1997) | | Box Office Market Shares (1998) | | Video Shipments Market Shares (1998) | |
|---|---|---|---|---|---|
| Feature films | 81% | 7 major studios | 87% | 7 majors studios | 83% |
| Sitcoms | 1 | Independents | 13 | Independents | 17 |
| Drama series | 1 | | | | |
| Children's | 12 | Total | 100% | Total | 100% |
| Sports | 3 | | | | |
| Other | 2 | | | | |
| Total | 100% | | | | |

*Note:* Data from (a) Media Dynamics; (b) and (c) Paul Kagan Associates.

overwhelming portion of revenues are generated by the theatrical feature films of major distributors. A small proportion of the feature films on video are direct-to-video movies, but these are often syndicated to cable or broadcast television. Most of the children's programs on video are also exhibited on cable or broadcast television. A very small proportion of video content relies solely on video rentals and sales for revenues.

Available data for cable television in the United States as reported in Table 5.2 is badly out of date, but also suggests the economic importance of multimarket syndication. Theatrical features dominate pay cable networks, and among the minority of originally produced programming on pay cable networks, including made-for-pay feature films, a large percentage of that is no doubt later released on video, on broadcast or basic cable channels. For basic cable networks, the proportions of originally produced programming are much greater. Aftermarket feature films and off-network programs accounted for less than half of viewing in this study, but for dramatic programming formats, the overwhelming portion of viewing was directed to aftermarket programming. The percentages of original dramatic and other entertainment programming on basic cable may have increased since the mid-1980s, but again, a substantial percentage of that programming later ends up on broadcast channels. Because the same basic economic forces

**TABLE 5.2**
***CABLE TELEVISION PROGRAM CONTENT BY SOURCE***
***1986***

| | All programming | Dramatic only |
|---|---|---|
| **Premium channels** | | |
| Original | 15% | 8% |
| Off-network | 2 | 2 |
| Theatrical film | 83 | 90 |
| Foreign acquisition | — | — |
| Total | 100% | 100% |
| **Basic channels** | | |
| Original | 56% | 3% |
| Off-network | 30 | 63 |
| Theatrical film | 12 | 32 |
| Foreign acquisition | 2 | 2 |

*Note:* Data from Waterman and Grant (1991).

are at work, there seems good reason to expect multi-market syndication business models also to dominate Internet television.

## CONCLUSIONS

The widely discussed opportunities for interactive and other new and innovative Internet entertainment, and for more narrowly focused and marginal programming in general, are backed by economic logic. Lower costs, virtually unlimited capacity, efficient interactivity, and more efficient business models will all contribute to making them possible. As the video and cable TV experiences suggest, however, there will also be powerful economic forces favoring relatively expensive, broad appeal programming (e.g., Hollywood movies) on Internet TV. By their nature, those are the types of programs that are most amenable to syndication on a variety of different media, as well as amenable to lucrative worldwide distribution.

A major challenge to the suppliers of Internet-original television programs will be to find lucrative aftermarkets for them, both in their domestic and foreign markets around the world. Inherently, these economic pressures also tend to encourage homogenization of content, as well as to limit the budgets of programs that have few alternative outlets. For example, interactive programming that depends on Internet architecture, or raunchier productions that do not adapt well to other media, will have major budget handicaps to overcome. It is a good guess that like basic cable TV, Internet TV programming will evolve into a dichotomous mix of niche-oriented, but relatively cheap Internet-original fare on the one side, and mass appeal, relatively expensive multimarket syndicated programming, on the other.

Few observers would claim that the diversity of entertainment programming, including much new and original, more sharply focused content, has not been greatly enriched by home video and cable television. But the results do seem to have fallen short of many of the hopes that visionaries' had for these media (spectacularly so in some cases, e.g., the grand hopes for "high culture" performing arts).

Historical experience suggests that these outcomes fell short of aspirations for three reasons. First, other things being equal, focusing program content on particular interests of small subsets of people seems to have stimulated demand (of both audiences and advertisers) less than was imagined. Second, visionaries underestimated the audience drawing power of high production values. By spending more on the best stars, locations, and special effects, and spreading those costs over a potentially very large multimedia audience, producers have been able to keep the lion's share of the viewers. Finally, many have underestimated the power of effective marketing. To support an opera in the United States certainly becomes more feasible with greater capacity (even if it can attract no more than 1% of the country's television homes), but an opera cannot compete with a blockbuster movie or a boxing match that can realize economies of scale in a national marketing campaign.

For some combination of all these reasons, its seems, the opera and other niche audiences have so far decided in the end to watch *Harry Potter*. So far, Internet-original entertainment programming shows great creative promise, but the economic challenges will not be easy to overcome.

## ACKNOWLEDGMENTS

An earlier version was published as D. Waterman (2001), The Economics of Internet Television: New Niches vs. Mass Audiences, *Info: The Journal of Policy, Regulation and Strategy for Communications, Information, and the Media*, Vol. 3, No. 3. I am especially indebted to A. Michael Noll, Andrew

Odlyzko, Ben Compaine, Robert LaRose, and to other conference participants for their comments, but they share no blame for remaining errors.

## REFERENCES

Adams, W., & J. Yellin (1976). Commodity bundling and the burden of monopoly. *Quarterly Journal of Economics*, *90*(3), 475–498.

Bakos, Y., & Brynjolfsson, E. (1999, December). Bundling Information goods: Pricing, profits, and efficiency. *Management Science*, *45*(12), 1613–1630.

Besen, S. (1986). Private copying, reproduction costs, and the supply of intellectual property. *Information Economics and Policy, 2*(1), 5–22.

Chetwynd , J. (2000, May 4). Hollywood experiments online with quantum leap to features. *USA Today,* p. 1D.

Clark, D. (2001, February 21). Napster to offer yearly fee to cd labels. *Wall Street Journal,* p. B6.

*The Economist* (2000, October 7). A Survey of E-Entertainment.

Einhorn, M. (2000, September 23–25). *Napster, copyright and markets.* Paper presented at the Telecommunications Policy Research Conference, Alexandria, VA.

Graser, Marc (2000, September 4–10). Only top-tier talent taps the till. *Variety,* p. 1.

Jackson, M. (2001, Spring). Using technology to circumvent the law: The DMCA's push to privatize copyright. *Hastings Communications and Entertainment Law Journal, 607.* Hastings College of the Law.

Katz, M. (1989). Home copying and its economic effects: An approach for analyzing the home copying survey. Report to OTA.

Kiernan, V., & Levy, M. (1999, Spring). Competition among broadcast-related Web sites. *Journal of Broadcasting and Electronic Media, 43*(2), pp. 271–279.

Konert, B. (2000). Broadcasting via the Internet: New Models of business and financing. *Trends in Communication, 7. Amsterdam: Boom Publishers.*

Long, Patrick (2000). Presentation at the "TV Over the Internet" conference, Columbia University, New York, November 10, 2000.

Macrovision, Inc. (1996). *Home taping in america.* The Second National Survey of VCR owners. *Summary Report.*

Mathews, A. W. (2000, December 18). Online providers of film, cartoons combine forces. *Wall Street Journal,* p. B1.

Media Dynamics (1999). *TV Dimensions.*

Motion Picture Association of American (2001). 2001 *Economic Review (www.mpaa.org).*

National Research Council (2000). *The digital dilemma: Intellectual property in the information age.* Computer Science and Telecommunications Board, National Academy Press.

Noam, E. (2000). *Will America be dominant?* Presentation at the "TV Over the Internet" conference, Columbia University, New York, November 10, 2000.

Noll, A. M. (1997). Internet pricing vs. reality. *Communications of the ACM, 40*(8).

Odlyzko, A. (2000). *From narrowband to broadband: Capacity requirements, architecture options and investment implications for the long distance network.* Presentation at the "TV Over the Internet" conference, Columbia University, New York, November 10, 2000.

Office of Technology Assessment (1989). *Home Copying Survey.*

Owen, Bruce M. (1999). *The Internet challenge to television.* Harvard University Press.

Picard, R. (2000). Changing business models of online content providers. *International Journal on Media Management, 2*(2), 60–68.

Samuelson, P. (1999). Intellectual property and the digital economy. *Berkeley Technology Law Journal, 14.*

Schlachter, E. (1997). The intellectual renaissance in cyberspace: Why copyright law could be unimportant on the Internet. *Berkeley Technology Law Journal, 12.*

Shaner, S. (1998). Relational flow and the World Wide Web: Conceptualizing the future of Web content. *Electronic Journal of Communication, 8*(2).

Shapiro, C., & Varian, H. (1999). *Information Rules,* Watertown, MA: Harvard Business School Press.

Snell, John (2001, April 30). Studios demand Internet services block access to pirated movie programs. *Knight-Ridder Tribune Business News.*

Tristam, Claire (2001, June). Broadband's coming attractions. *Technology Review.*

Varian, H. (1989). Price discrimination. In R. Schmalensee and R. Willig (Eds.), *Handbook of industrial organization.* New York: North Holland.

Waterman, D., & Grant, A. (1991). Cable television as an aftermarket. *Journal of Broadcasting and Electronic Media, 35*(2), 197–188.

Waterman, D., & Yan, Z. (1999, Fall). Cable advertising and the future of basic cable networking. *Journal of Broadcasting and Electronic Media, 43*(4), 645–658.

# 6

# Broadcasters' Internet Engagement: From Being Present to Becoming Successful

**Bertram Konert**
*European Institute for the Media*

Digitization and globalization, deregulation of the telecommunications sector, opening up the broadcast sector to privately owned companies, and the emergence of new cross-media companies are transforming traditional broadcasting. Developments in technology, in particular, are the driving forces behind the transformation of media and communications. This analysis of Internet broadcasting will take into account both socio-economic and technological conditions.

## THE SOCIOECONOMIC AND TECHNOLOGICAL CONTEXT

Changing market conditions challenge the broadcasting sector's role and field of activity. Global players are investing heavily in information and communications technology (ICT) and the new media high potential growth markets. The economic significance of information technology and telecommunications is evident from the growing proportion of gross domestic product (GDP) that these markets represent. The turnover on ICT in 2000 as a percentage of GDP was 6.3% in Western Europe (4.9% in 1997) and 8.7% in the United States (7% in 1997) which shows the overall increase in the contribution of this sector to the national economics (FVIT, 1998; BITKOM, 2001).

The extraordinary economic significance of information and communications technology is also due to the fact that ICT is an innovative multifunctional technology enabling other products and services and representing a high proportion of the potential value creation in other markets. Increased process and product innovation and the ongoing convergence between ICT and media serve to reinforce the increasing significance of multimedia-related markets for overall economic development.

At the heart of the multimedia development is the Internet. The fact that the Internet is developing into a new vehicle for business to business (B2B), business to consumer (B2C), consumer to business (C2B), and consumer to consumer (C2C) communication has been seen as a major catalyst for achieving new profits in "old" and "new" economic sectors alike (A Survey of E-Commerce, 2000).

About 373 million people worldwide used the Internet on a regular basis in 2000. This number increased by 59% from 1999 (BITKOM, 2001). It is extremely difficult to measure the number of Internet users directly, because several people may use one individual account (business and private), and figures from Internet service providers (ISP) and online services cannot be verified with any degree of certainty. The increase in the number of Internet hosts between 1999 and 2000 in Europe and the United States provides a more reliable picture of the growing importance and intensity of Internet use in individual countries.

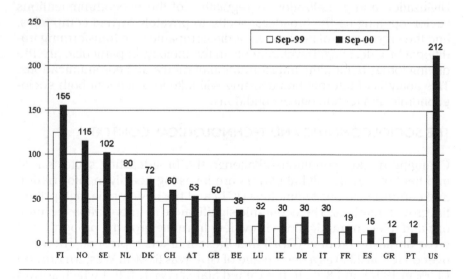

FIG. 6.1.   Internet Hosts* per 1,000 Inhabitants in Western Europe and the United States. From EITO (2001, p. 488).

The growth rates are very impressive and illustrate the large lead that Scandinavian countries have gained over other European countries. However, even in the Southern European countries where the Internet use and ICT investments have traditionally been at a lower level, the number of Internet hosts per 1000 inhabitants has increased more rapidly (see also Konert, 1999a).

In addition, the sociodemographic structure of Internet users gradually adjusts to the average population. Unlike the early years when Internet users tended to consist almost exclusively of young men with high incomes and high education, today women, the older generation, and people with average incomes and education are using the Internet more frequently. The Internet, especially in developed countries, is on its way to becoming a mass communication infrastructure.

So, what are the major reasons for the huge success of the Internet? Decisive "success factors," especially of the World Wide Web (WWW), include a breaking up of geographical limitations and time delay, improvements in accessibility and user-friendliness, new services and content with "added value" to the user, interactive communication with direct links and feedback between supplier and recipient, and integration of multimedia applications. The increasing use of the World Wide Web in the business and entertainment spheres poses new challenges to traditional broadcasters. To meet this challenge, broadcasters will have to become more actively involved in new forms of media, rather than to rely on their mere presence.

A heated debate is currently underway concerning whether and to what extent TV and PC transmission methods, services, and terminals might fully merge in the future. There can be little doubt that the two areas will continue to move closer with increasingly combining functions, although differences in users' preferences, social and cultural factors will slow down the trend toward a complete integration that could supplant existing systems. At present, Internet TV is used above all to present additional special interest niche programs, video-on-demand services, and complementary program-related offers.

When dealing with the relation between technology and society, cultural conclusions are usually derived from technological premises based on the assumption that social developments somehow follow technological achievements, and more or less adapt themselves to technological developments.

This view overlooks the influence of social and cultural factors on the development of technological innovations, on their success, and on the speed of their diffusion. The origin and shaping of new technology is essentially influenced by cultural models of its usage, and by cultural patterns of perception (Rammert, 1996). For a number of years, there has been an ongoing discussion about the individualization and multiplication of lifestyles, and the fragmentation and segmentation of social groups. Fragmentation means also a decoupling of time structures in the world in which people live and work (more flexible working times, more weekend work, etc.) and a change in the distribution of leisure time throughout the day and week. These developments help create a demand for information and entertainment independent of time and place while intensifying the fight for recipients' attention and limited time.

## NEW COMPETITIVE ENVIRONMENT

Traditional broadcasters understand that the increase in digital programs and the Internet have opened a new competitive environment. As consumption time for media use is limited, traditional broadcasters fear the popularity of the Internet will have a negative effect in the long term on the number of people watching television.

Currently, hundreds of TV stations are already active on the Internet. The database of the "World Wide Internet TV" portal, for example, in May 2001 contained nearly 400 TV stations worldwide with live or recorded programs via the Internet.[1] However, at present this portal has only 73 programs listed with a bandwidth from 100 Kb/s up to 300 Kb/s for a better TV/video quality.

With this form of Internet broadcasting, individuals can select special programs and channels from the web as a service available "on demand." In fact, rather than having to download an audiovisual file completely to be played later on a multimedia PC, people can now receive sound and video images right from the beginning of the downloading process using software such as RealPlayer (RealNetworks), MediaPlayer (Microsoft), or Quicktime (Apple). This is known as streaming technology. At present, live or on-demand Internet TV services still have a poor quality due to the relatively small bandwidth of Internet access via normal telephone lines. However, increasing broadband high-speed data transfer rates will allow providers to offer interactive communication with TV quality. Future broadband delivery systems will be realized with competitive technological systems. At present, the most likely candidates to succeed are asymmetric digital subscriber line (ADSL), the digital transmission technology through copper telephone cable; digital TV cable (DVB-C), satellite (DVB-S), or terrestrial systems (DVB-T) for the transmission of audiovisual and data services according to digital video broadcast (DVB) standardization; and universal mobile telecommunications systems (UMTS), the future technology for mobile services.

There are three categories of Internet broadcasters competing on the supply side: *Traditional public service broadcasters* present their pro-

---

[1]http://wwitv.com

grams on the Internet along with new services and supplement their TV programs with specific additional services. *Commercial broadcasters* look for new opportunities to strengthen their customer relationship with additional and new services, particularly for special target groups. *New (cross-media) actors and producers* have, to date, avoided the technical, financial, and administrative expense of setting up their own broadcasting system. In this commercial context, which sees the players entering high-potential growth markets, an increasing integration of business activities (mergers, strategic alliances) at various stages of the value chain can be seen. Furthermore, the process of convergence in the media and communications sectors, accelerated by the rapid advance of the Internet, is stimulating far-reaching changes in the traditional stages of the media value chain (Fig. 6.2). The traditional media value chain is eroding as new opportunities present themselves to suppliers to create direct relationship and access to their customers from each point of the networking process. This creates new opportunities to produce, arrange, and post programs and services directly online circumventing traditional broadcast actors or authorities and without having to rely on traditional "intermediation."

These changes allow new companies to produce and distribute new broadcast services directly for special interest groups via the Internet with a comparatively low budget. For example, the European Internet TV company Canalweb[2] aims at special interest television with a highly targeted

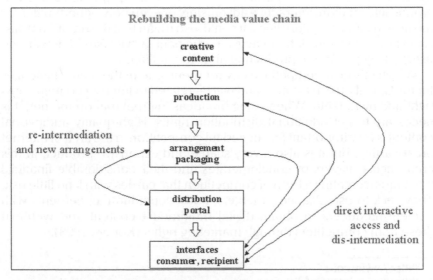

FIG. 6.2.   Rebuilding the media value chain.

[2]http://www.canalweb.de

content ("narrowcast approach") and at fully interactive TV services via on demand services and interactive program guides (IPG).

New "re-intermediation" suppliers that select, arrange, and combine the audiovisual products and programs will be more important for those Internet broadcasters or producers whose brand name is not well known enough for them to attract an audience by themselves. At present, nearly all well-known Internet portals like Yahoo, MSN, AOL, or T-online offer special links to TV, film, and video resources. Furthermore, completely new portals like "WorldWide Internet TV"[3] and Internet TV List[4] (especially for world-wide TV programs) are active in the "re-intermediation" of Internet TV.

Internet TV content is not presently comparable to regular TV. Short films, documentations, news, or soaps are regarded as successful Internet TV formats, limited to a length approximately between 1 and 6 minutes (Becker, 2000; Hagen, 2000). MTV´s video clips especially inspire the producers of so-called Web-clips distributed via Internet. Due to the limited bandwidth and the different situation of Internet usage, producers of Internet TV must pay special attention to the time element. In the United States, for example, the term "Break TV" was coined to describe this format, because many viewers watch Internet TV and film clips during commercial breaks. Two examples of special short film portals are "The Sync"[5] and "absolutFilm,"[6] which offer video clips of various bandwidth speeds via the Internet.

Furthermore, interactive use will be more important in the future. Internet TV opens new possibilities to present complimentary background information, interactive advertising, or even new opportunities for the user to influence story lines. It could also mean that the traditional passive viewer develops into an active participant, a "viewser" (viewer and user) taking part in the production and story telling.

Within this new competitive environment, one of the main challenges for traditional broadcasters is and will be related to Internet content copyright and ownership. When using broadcasting material on the net, it is necessary to get additional distribution rights. High-quality audiovisual content is very important for successful competition in the area of Internet broadcasting, but it is also a very scarce and expensive product. In this area, media players or conglomerates and their considerable financial backing are creating a form of competition that carries with it no little risk: They seek to use alliances, mergers, and cooperation agreements with content producers to obtain digital multimedia content and withhold these from competitors through marketing rights (Konert, 1998).

---

[3]http://wwitv.com
[4]http://www.internettvlist.com
[5]http://www.thesync.com
[6]http://www.absolutfilm.de

## STRATEGIC OBJECTIVES

The fight for consumers' time and attention, that valuable and rare commodity, will continue to intensify. Because traditional TV viewer rates are falling as Internet use increases, there is an urgent need for compensation strategies. In principle, there are three strategic objectives broadcasters attempt to realize online: First, they have to secure their future market share; second, they should transfer their brand images and traditional core competencies to Internet services; and, third, they should strengthen an online position that distinguishes them from competitors.

### Securing Future Market Share in the Long Term

Securing market share in the long term is one of the main challenges facing traditional broadcasters, even if their activities in the new media market are "cannibalizing" their traditional business (European Communication Council Report, 1999, p. 178). As the Internet will grow and not only become a mass communication network (which it not yet is) but also a competing media form for new audiovisual and TV-oriented services, traditional broadcasters have to secure their future. This is only possible if they take part in technical up-to-date developments in the area of new media and if they acquire the expertise and experience needed to be successful and competitive.

### Transfer of Brand Images and Core Competencies

The transfer of brand images and core competencies is especially necessary in this relatively unstructured and intricate Internet environment as users frequently chose brand names and services familiar to them from other areas. Companies compete with other web portals, but the effectiveness of broadcast programs makes it easier to position their brands on the Internet.

Broadcasters who are, for example, renowned for their journalistic professionalism and/or viewer-oriented programs have a definite competitive advantage even in the area of new media. One example of this is BBC Online. BBC Online is the most popular site in Britain and the most visited nonportal Web site outside the United States.[7] In the United Kingdom, the Web site receives approximately 200 million page impressions every month. Latest figures give it a 25% reach, which means one quarter of the Internet-using population in the United Kingdom visits this site (Cozens, 2000).

---

[7]Frankfurter Allgemeine Zeitung, February 15, 1999.

## Strengthening of Distinctive Online Position

Broadcasters have to strengthen their online position and make sure they distinguish themselves from their competitors. Adapting technological up-to-date developments, launching online services with a specific "added value" or offering thematic contents for target groups are all ways to establish a distinctive online position. New forms of "mass customization" and personalized portal sites (see chap. 3) are used to launch efficient and competitive services. Although the realization of competitive advantages related to technological developments and new services may be successful in the short term, it is less likely to prove a success in the longer term. In this fiercely competitive environment, companies would do well to build up a unique sales proposition (USP) in their online presence. For public service broadcasters, this USP is to be found in their public service mission. Public service broadcasters' Internet engagements could compensate for some Internet specific disadvantages such as lack of filters, reliability, and orientation for the general public in an intricate environment by concentrating on their core tasks and transferring those to the online area.

## INSTRUMENTS

### Exploitation of Internet Specific Success Factors

Broadcasters should not only rely on their brand name and core competencies to achieve online success. They have to anticipate and exploit the main "success factors" of the Internet if they are to realize their strategic objectives. Comparative research has shown that the following criteria should be taken into consideration to get sustainable attention and visits by Internet users (Konert, 1999b).

*Up-to-Dateness.*   One of the main advantages of the Internet over more traditional media is its immediate access. Other media forms (e.g., newspapers) are constantly faced with geographical limitations and time delays. The Internet is potentially faster and more up-to-date. Having said that, it is vitally important that Web sites be kept up-to-date to provide Internet users an incentive to visit the site more frequently.

*Content Presented.*   Content is "king," and web pages that provide the user with real added value receive higher attention and will be sought out repeatedly as reference points. Internet broadcasting content can enrich programs with additional background information and new services for special target groups, not least by using interactive multimedia capacities. Furthermore, on-demand access to digitized program archives at all times appeals to nearly all online users. The resources and the expertise in presenting high quality audiovisual material are among the main advantages traditional TV and radio broadcasters have over their online competitors.

*Interaction.*   Being online offers new possibilities for broadcasters to intensify their links and communication with their audience. Meanwhile, most suppliers offer interactive services, including e-mail, news, chat-rooms, and guest books. Those areas related to specific TV programs are particularly well suited for interactive communication due to a common thematic agenda (e.g., soaps, documentation, adviser, and serial programs). However, the success and quality of these services are dependent on the level of interaction between suppliers and users.

*Presentation.*   Functionality of the technical interface has been adapted to users' needs. Clarity, consistency, transparency, and easy access are the main requirements for clearly arranged web pages and user-friendliness.

*Multimedia.*   The use of new multimedia technologies expands the broadcasters' ability to shape and present complementary or new services. Digital recording and playing of video and audio streams (live and/or on-demand), and the integration of animated graphics within complex issues are examples of ways in which new multimedia possibilities make the online offers more attractive. The use of broadband transmission will increase the technical quality of video streaming and thus respond to users' interests. Internet TV content will not be the same as present-day TV programs.

## Promoting Synergy Between TV and Internet Services

Broadcasters must use synergy. TV programs use their brand name to draw viewers' attention to supplementary Internet services, and their Internet activities add additional value to their TV programs.

Endemols' *Big Brother,* a reality TV program shown in Europe and the United States, is an example of how this kind of synergy between the traditional and new forms of media can work. The German *Big Brother* Web site (Fig. 6.3) offered viewers the opportunity to stay up-to-date around the clock, giving them access to webcams all over the *Big Brother* compound (including infrared cams for the sleeping rooms).

By providing news and stories about the candidates and their environments, on-demand archives, chat-rooms, merchandising, and voting opportunities, online services leverage synergy with the TV programs. As a result, visits to the RTL Web site increased by about 30% between December 1999 and May 2000.

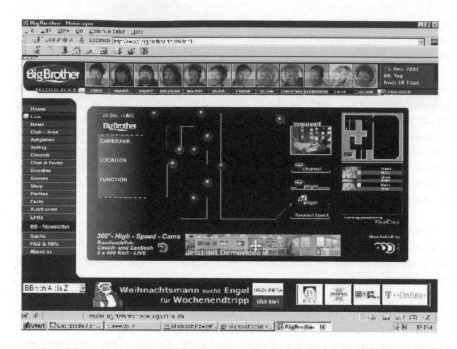

FIG. 6.3.   Synergy between TV and the Internet: *Big Brother* home page.


Furthermore, combining television and the Internet opens up new ways of reusing existing digitized content. Broadcasters like the BBC, CNN, or ARD have at their disposal huge resources of video material, which can be delivered over the Internet. The German broadcaster ARD, for example, offers an online edition of its main news program "tagesschau,"[8] which is a very successful on-demand service in Germany. Copyright, ownership, and the costs of additional distribution rights are key concerns when repurposing content via Internet.

### Localized Content Via Global Media

By its very nature, the Internet is a global medium. But, as with traditional television stations, the Internet broadcast services require a strong emphasis on local and regional content. Content providers that do not lose

---

[8]http://www.tagesschau.de

sight of the regional context in which people exist have a better chance of launching interactive and lively services.

Even in well-developed information societies, from 80% to 90% of public communication services will continue to be domestic or national (van der Meulen, 1999). This implies that coverage of domestic news, events, and regional services will remain important. The CNN online service, "CNN interactive," for example, in expanding regional services spreads into Europe, Africa, and Middle East with local language sites and strong local partnerships.

Furthermore, growing regionalization applies to commercial activities where trust, for example, is an important aspect of e-commerce. With regard to points of payment and reclamation, regional Internet suppliers have a better chance than international suppliers.

## Mass Customization

Thus far, small special interest groups have not been economically interesting to traditional TV and media companies. A different cost structure, however, means that Internet services can target smaller user groups far more cost-effectively (Goldhammer & Zerdick, 1999). Thematic channels or special interest portals offer new possibilities to meet users' specific needs. The mechanism of "mass customization" can be used to increase users' loyalty to specific Web sites and can facilitate target advertising and launch special pay services (see chap. 5).

## Personalized Portal Sites

Successful portal sites such as Yahoo! or AOL offer selective and structured services and provide their operators with lucrative advertising opportunities. The midterm strategy of broadcasters aims at setting up portal sites to bring more traffic to their Web sites. However, portal sites with a huge variety of content and services make it more difficult for users to find the content they are really interested in quickly and easily. Therefore, the installation of special interest portals or personalized portal sites with more possibilities for individual arrangements will be increasingly important in the future.

CNN.com,[9] for example, offers a personalized service called "myCNN," which allows users to select specific news areas and services (e.g., sports, world news, lifestyle, etc.) and use them to create their personal CNN portal sites (Fig. 6.4). Access to these personalized portal sites is organized via password and user name.

---

[9]http://www.cnn.com

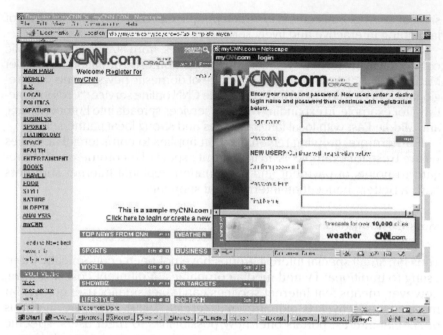

FIG. 6.4.  Personalized portal sites: "mycnn" homepage.

## NEW REQUIREMENTS

In order to achieve the strategic objectives and use the online instruments effectively, broadcasters must understand the requirements for success in this new media market.

### Cross-Media Activities and Partnerships

Broadcasters need competent partners to obtain online expertise. The German public service broadcaster Zweites Deutsches Fernsehen (ZDF), for example, started a strategic partnership with Microsoft as early as 1996 to get technical support for the launch of its Web site "www.zdf.de" (Jarras, 1997). Cross-sector partners profit from broadcasters' brand names and high-quality content to generate online traffic. In March 2001, the German Internet provider T-Online concluded a new partnership with the ZDF to use the content of the news program "heute" exclusively on its portal by paying a licensing fee. As a result, the corresponding web address "heute.t-online.de" will be shown in the respective news program (T-Online Wählt ZDF als Nachrichtenpartner, 2001). This new partnership brings the previous cooperation between ZDF, Microsoft, and NBC (zdf.msnbc.de) to an end.

Cross-media cooperation is also occurring between commercial TV companies and newspaper publishers. Examples of this are N24[10] (cooperating with Frankfurter Allgemeine Zeitung) and n-tv[11] (cooperating with Handelsblatt). This type of cross-media cooperation poses new questions about possible and potentially problematic joint ventures and alliances, particularly for the public service mission of public service broadcasters.

## Organizational Restructuring

Technical restructuring measures are not implemented in isolation. Embedded in internal organizational changes, broadcasters' online success depends on a professional and independent business organization that relies on cross-subsidization as little as possible. Broadcasters need to become more flexible, and to speed up their decision-making processes to compete in this highly competitive environment (BBC, 2000). For that reason, broadcasters have started to establish joint business units, where all Internet and new media activities are assembled. These trends can be seen within commercial as well as public service broadcasters. Public service broadcasters try to establish an online presence in addition to their traditional activities (third pillar beside television and radio) and commercial broadcasters particularly set up new affiliated companies responsible for all Internet and new media activities (e.g., Kirch New Media, RTL New Media).

## Skills

Broadcasters have to make sure that they have the skills needed to compete effectively in the digital world. Growth in new services and programs at this level highlights, for example, the need for professionalism, effectiveness, and journalistic reliability. The expansion of traditional broadcasting to interactive and multimedia online services presents new challenges for those working in this sector on two levels: First, they require new Internet technologies and online material to get the information that they need for their daily work. Second, they have to know how to use these new technologies and sources for additional services offered to Internet users and recipients. This applies not only to technical knowledge, but also to the presentation of online services and the interaction with recipients before, during and after programs (e.g., via e-mail, chatrooms etc.).

---

[10]http://www.n24.de
[11]http://www.n-tv.de

## MODELS OF FINANCING

This chapter discusses the main question of the source from which the money to finance the online strategies, the instruments, and the new requirements will come. It will not be possible for broadcasters in the long run to subsidize business activities in the area of new media with revenues from their traditional activities. Broadcasters must strive for online success, not mere presence defined as web page impressions and visits, but as return on investment (ROI).

Internet technology promises for new revenue streams. New revenue models (Fig. 6.5) include direct and indirect proceeds. Direct proceeds are revenues that are paid directly by the customer to the supplier. Indirect proceeds are revenues that are mainly paid by third parties or the general public. This chapter outlines the main possibilities and combinations of these proceeds to increase revenues for Internet broadcasting.

### Direct Proceeds

*Transaction (Pay-Per-Use).*    As is the case in digital television, pay-per-view services offer suppliers the opportunity to increase their sources of

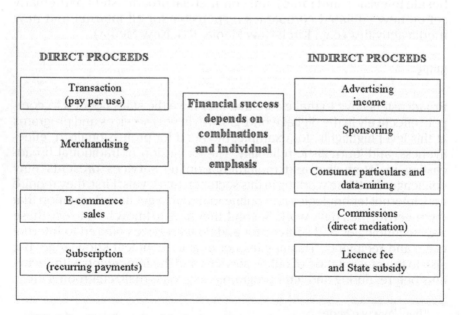

FIG. 6.5.    Revenue models for broadcasters' online activities.

revenue. It is made technologically possible by the direct link online broadcasters have with their recipients via back channel (point to point). The impetus for offering pay services comes from an interest in serving specific niche markets of those users willing to pay a premium for high-end services or specific content.

Pay-per-use models are not new in the online area. They are currently being used by some newspaper publishers, for example, that offer online archive search services, whereby the user has to pay for each article found. Similarly, Internet broadcasters can offer special interest content (e.g., documentaries, film festivals, live concerts) or the access to their video and audio archives as an exclusive pay-per-use service. Other models of financing include access time-related extra payments for special high quality services (e.g., online games). It is, however, far from easy to persuade users to pay for these services, as long as they can access similar services elsewhere free of charge. This situation would presumably change if a supplier had "exclusive" content that could not be accessed from other sites free of charge.

Presently, public service broadcasters, who get their income mainly from license fees or state subsidies, are largely prohibited from offering pay-per-use services. Such services, aimed as they usually are at special target groups, contradict their societal core tasks of free access for the general public. In addition, public service broadcasters face considerable resistance from commercial broadcasters and politicians, who fear unfair competition.

*Merchandising.*    Online merchandising is a potential additional main source of revenue. Merchandising products, offered by broadcasters over the Internet, are generally specific to programs or to the broadcasting station itself. These products, such as coffee mugs, T-shirts, caps, videos, books, and CDs can be ordered online. Figure 6.6 shows the BBC shop, offering merchandising products such as videos, games, and books related to the successful children program *Teletubbies*. In addition to increasing companies' sales revenues, merchandising helps strengthen the brand name of the program or station in question.

*E-Commerce Sales.*    Another way online broadcasters can increase revenues is through expanded e-commerce sales that are not directly linked to their programs. E-commerce sales concern not only products such as books, videos, or CDs, but also other cross-marketing products or services, including electronics and travel services. Figure 6.7 shows a clipping from the commercial German broadcaster ProSieben (http://www.prosieben-shop.de), offering, for example, computer games, modems, and telephones, all of which can be ordered online. At present there is no data available on the commercial acceptance of these new services.

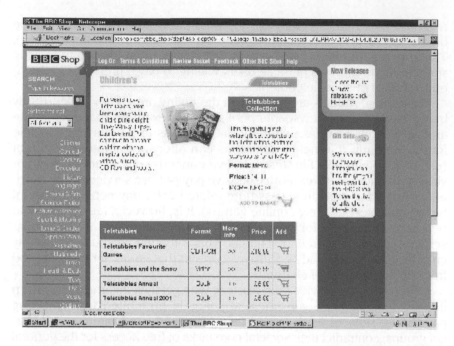

FIG. 6.6.    The BBC shop online.

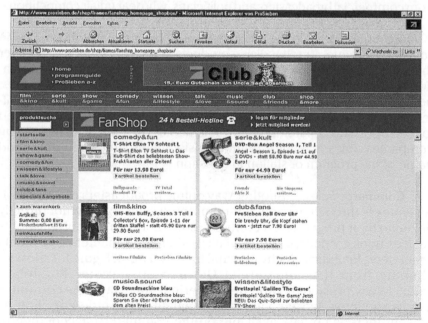

FIG. 6.7.    The "Pro 7 Club" online shop.

It is important that the back office functions required for the e-commerce are handled efficiently. Broadcasters usually work with competent partners from other areas, who know the online business well and are equipped with the appropriate technological and personal infrastructure for processing volume online sales. Of course, transaction and payment systems must be easy to use, transparent, and secure against fraud.

*Subscription.* Subscriptions as recurring payments for the access to special online services could be one of the main sources of generating proceeds for online business. As in the case of pay-per-use models, commercial success depends on users' willingness to pay for the services. Users must have a compelling interest in particular content that motivated them to pay for it when they are accustomed to getting online content for free. At present, this revenue model is mainly successful with professional business services such as online databases, and can be compared to digital TV subscriptions in the areas of sports and erotica (European Communication Council Report, 1999, p. 171). A current attempt to realize an efficient subscription-based service on the Internet is the music file exchange system offered by Bertelsmann and Napster.

Internet broadcasters who are primarily general interest suppliers may find it difficult to be successful with this model of financing in a highly competitive environment. In addition to a more content-oriented subscription model, broadcasters might attempt to launch new services through partnerships with telecommunication companies and Internet service providers, who already use subscription models such as flat rates and basic fees.

### Indirect Proceeds

*Advertising.* Thus far, viewer rates and page visits are the main criteria used to measure the success and attractiveness of online advertising in-

FIG. 6.8. Online advertising revenue in the United States: Q2–Q3 2000 (in millions).

vestments. Third-party advertising is a leading source of revenue. An increase in advertising revenues for online broadcasters largely depends on the economy. Advertising revenues will continue to be one of the main revenue sources for online services in the near future. The Internet Advertising Bureau (IAB) reported total online revenues for the year 2000 of $ 8.2 billion in the United States (Interactive Advertising Bureau (IAB) Reports $8.2 Billion Online Ad Revenue in the United States, 2001).

But this hypergrowth might be over. The Internet Ad Revenue Report shows that online advertising revenue decreased slightly, by 6.5%, between second quarter and third quarter 2000. According to IAB and PricewaterhouseCoopers, this decrease from Q2 to Q3 and the comparable lower increase with historical levels of 9% to Q4 2000 ($ 2.2 billion) are a reflective of the overall slowdown in ad revenue across all media sectors. Banner advertisements continue to be the dominant form of online advertising with 47% of overall online advertising in 2000. According to these current figures, relying solely on advertising revenues is not a good long-term strategy.

Many Internet users find advertising banners a nuisance because they tend to slow download times and their content is unwanted. The latest figures suggest that click-through rates for banner ads are dropping (Markt für Online-Werbung wächst auf 300 Millionen DM). Advertising companies try to solve these problems with special interest advertising, adapting to the unique characteristics of online users. To do this, they need to develop individual user profiles and more "intelligent" interactive banner ads. These developments make Internet privacy a paramount concern.

The importance of banner advertisement revenues will cause problems, especially for those public service broadcasters who are not allowed to generate this kind of income. Politically speaking, online advertising is viewed more critically in European countries than television advertising. For example, according to "ARD- and ZDF-Staatsverträge," public service broadcasters in Germany have been explicitly prohibited form engaging in online advertising since April 2000.[12] In the United Kingdom, this political dilemma has brought about a situation in which the BBC has two different online presentations: the "public service site,"[13] which does not use advertising, and the e-commerce-oriented site "beeb.com"[14] by BBC Worldwide, which does use it.

Public service and commercial broadcasters with a professional journalistic mission should distinguish between commercial and noncommercial items in the same way this distinction is made between program and advertising on TV. That this is not always the case is visible, for exam-

---

[12]http://www.artikel5.de/gesetze/rstv-4.html
[13]http://www.bbc.co.uk
[14]http://www.webguide.beeb.com

ple, on the commercial Web sites of CNN[15] and RTL,[16] neither of which clearly marks the web banners as advertising.

*Sponsoring.* On the one hand, sponsorship is being used as an additional source of income for online services. On the other hand, well-chosen sponsorship might give the sponsors a higher profile than normal banner advertisements would. Sponsors hope that users will more closely associate online services and content with the advertiser. Sponsored online activities could be related, for example, to special chat meetings with stars from entertainment and sports or to live streaming events. In 1999, the German Telecommunications Company T-Online sponsored, for example, the launch of the German "n-tv" on-air broadcasting program via the Internet. Like event sponsorships, support of well-chosen online activities might help to heighten brand name awareness more efficiently than other kinds of advertising. Sponsoring is the second most important source of online advertising revenue (after banner advertising), accounting for 27% of the $4.6 billion in advertising revenues in the United States over 1999.[17]

Some public service broadcasters are faced with the same online advertising dilemma. In Germany, advertising and sponsoring is explicitly forbidden as an additional source of income for public service broadcasters' online activities. The theory is that advertising and sponsorship revenues supporting PSB Web sites would be unfair competition for commercial Web sites.

In addition, there is a risk that income from sponsorship might alter the balance of online services by putting pressure on public service broadcasters to offer more services that are attractive to sponsors. Nevertheless, there are obvious gray areas in online advertising. Is the reference to streaming software (e.g., Microsoft's media player or RealNetworks' RealPlayer) advertising or sponsoring, or is it merely a technical necessity? And what is to be made of a Web site telling its users that it is "Best viewed with Internet Explorer version x.x"? In addition to the problem of ad marking, these few examples show that more clarity and harmonization in the field of online advertising and sponsorship is needed throughout Europe, particularly in the area of broadcasting on the Internet.

*Consumer Particulars and Data-Mining.* Online marketers are extremely interested in high quality demographics allowing them to target specific user groups. Advertising partners are interested in page visits, effective user time, and individual users' sociodemographic data. In addition to

---

[15]http://www.cnn.com
[16]http://www.rtl.com
[17]http://www.iab.net

traditional television measurement methods, such as Gesellschaft für Konsumforschung (GfK) in Germany, two rival agencies, Media Metrix Inc. and Net Ratings Inc. in the United States, are developing spot-checks of user behavior on the net (Messung der Internet-Nutzung ist umstritten, 1999).

With today's sophisticated Internet technology companies have an efficient instrument to collect, store, and analyze online activities with user profile data. For broadcasters, this data is very useful. When users introduce themselves providing their name and address, personal data can be used to present detailed user profiles to sponsors and advertisers, for target advertisements. The potential for providing valuable tailor-made services that can be offered, observed, and changed quickly is rapidly increasing.

Furthermore, this valuable information can be sold to interested third parties. The mere fact that companies are able to make a profit using very specific user-related information means there is a major privacy issue. At present, there is a privacy debate being held in Europe, as well as the "safe harbor" dialogue with the United States. It is uncertain what impact these developments will have on the growth of online advertising and users' online behavior. In principle, it should be guaranteed that without users' explicit agreements, systems are not allowed to collect information that can be traced back to individual users. This is often framed as an "opt in" versus an "opt out" debate. The issue is whether consumers are protected from privacy violations unless they voluntarily provide (i.e., opt in) this information or whether information can be collected unless a consumer specifically state (i.e., opt out) that they do not wish them to do so.

*Commissions.*     Commissions or affiliate programs seek a new revenue model to integrate advertising concepts and participation in the sales revenues (Goldhammer & Zerdick, 1999). The concept is based on a pay-per-sale mechanism, in which a Web site supplier, a broadcaster for example, places his business partner's banner on the broadcaster's site and receives a fixed percent share for each product sold by the business partner through the banner. In this way, Internet broadcasters can increase their income, and sellers can increase their e-commerce customer base. The online bookstore Amazon.com and the CD seller CD-Now are pioneers in this area, paying revenue splits between 3% and 15%. For broadcasters, this model is particularly appealing as special programs have the potential to generate additional buying incentives for customers linked via banners to appropriate sellers.

*License Fee and State Subsidy.*     Public service broadcasters in Europe are primarily financed by a mixed system of license fee, direct subsidies from the state and advertising (European Institute for the Media, 1998). The main difference between public service broadcasters who rely mainly on license fees and those who rely on direct government grants is that the former enjoys the more stable and predictable system of license fee funding (McKinsey & Company, 1999).

Due to commercial competition with private broadcasters, the increasing costs of purchased programs (e.g., films and sport), and the high investments for digital technologies and online services, the economic situation of public service broadcaster is becoming ever more difficult. Political initiatives and measures to increase license fees and state subsidies are almost universally unpopular, especially among those who prefer private broadcast services (TV and Internet). Public service broadcasters realize that their traditional sources of income will gradually disappear. However, if public broadcasters are not able to attract younger audiences, potentially with complementary online services, their public service role may be compromised in the long run. As already stated, public service broadcasters have to become more actively involved online. This means that they have to find a way to make their online activities commercially viable. This can be achieved through traditional means such as license fees or public funding. If these prove insufficient, public service broadcasters should be allowed to exploit other ways of financing their online activities. Having said that, these alternative finance models could call into question the public service broadcasters' very legitimacy.

## CONCLUSIONS

The increasing use of the World Wide Web in the business and entertainment spheres poses new challenges for traditional broadcasters. To meet these challenges, broadcasters will have to become more actively involved in using the new forms of media. These new online engagements must generate their own revenues and produce profits. It is not possible in the long term for new media business activities to be subsidized with revenue resources from traditional broadcast activities. Traditional broadcasters' Internet engagements must succeed not only in terms of web page impression numbers and visits, but more importantly, in terms of return on investment (ROI). In principle, it will not be possible for broadcasters to rely solely on one or two sources of revenue. Should hypergrowth decrease, as seen in the third and fourth quarters of the year 2000, then broadcasters' online investments and activities would be endangered. In the long run, financial success will depend on broadcasters' abilities to combine various new and lucrative direct and indirect proceeds from new media engagements.

## REFERENCES

A survey of e-commerce. (2000, February 26). *The Economist.*
BBC. (2000). *Building One BBC. Organising for the future.* Report of the BBC director. Source: http://www.bbc.co.uk/info/bbc/pdf/onebbc.pdf

Becker, P. (2000). Web-clips und web-soaps: Die Inhalte des Internet-TV. [Web-clips and web-soaps: the content of Internet TV]. *c't 2000, 13.*

BITKOM (Bundesverband Informationswirtschaft, Telekommunikation und Neue Medien). (2001). Wege in die Informationsgesellschaft. [On the Way to the Information Society]. Status Quo and Perspectiven Deutschlands im internationalen Vergleich. [Status quo and perspectives on Germany in an international comparison]. Retrieved from http://www.bitkom.org.

Cozens, C. (2000, October 31). BBC Online—What you need to know. *The Guardian.*

European Communication Council Report. (1999). *Die Internet-Ökonomie: Strategien für die digitale Wirtschaft* [Internet economy: Strategies for digital business].

European Institute for the Media. (1998). Perspectives of public service television in Europe. *Media Monograph, 24,* p. 44.

(FVIT) Fachverband Informationstechnik. (1998, June 1). Wege in die Informationsgesellschaft. [On the Way to the Information Society]. *Beilage zum Medienspiegel, 22(23).*

Goldhammer, K., & Zerdick, A. (1999). Rundfunk online: Entwicklung und Perspektiven des Internets für Hörfunk *und Fernsehanbieter [Online broadcasting: Development and perspectives of the Internet for radio and TV broadcasting].*

Hagen, Y. (2000, June 16). Internetsender: Die schleichende Revolution in der Filmindustrie [Internet broadcasting: The creeping revolution in the film industry]. *Süddeutsche Zeitung, 137.*

Interactive Advertising Bureau (IAB) Reports $8.2 billion online ad revenue in the United States. (2001, April 23). *IAB Press Releases.* Source: http://www.iab.net

Jarras, H. D. (1997). Online-Dienste und Funktionsbereich des Zweiten Deutschen Fernsehens [Online services and performance sector of the Zweites Deutsche Fernsehen]. *ZDF-Schriftenreihe, 53.*

Konert, B. (1998). Economics of Convergence, Part 1 and Part 2, *Bulletin of the EBU Strategic Information Service,* No. 11, April 1998, and No. 12, May 1998.

Konert, B. (1999a). ICT and multimedia in Western Europe and North America. *World Communication and Information Report (WCIR) 1999–2000.* UNESCO Publishing.

Konert, B. (1999b). Expertise zur ARD-domain www.ard.de [Expert opinion referring to the ARD domain www.ard.de]. *Strategische Optionen zur Realisierung von nachhaltigen Qualitätsmerkmalen im Online-Bereich* [Strategic Options for the Realization of Sustainable Quality in the Online Sector]. Düsseldorf: EIM Expertise.

Markt für Online-Werbung wächst auf 300 Millionen DM [Market for online advertising grows to 300 Million DM]. (1999, December 16). *Frankfurter Allgemeine Zeitung.*

McKinsey & Company. (1999). *Public service broadcasters around the world* (p. 30).

Messung der Internet-Nutzung ist umstritten [Measuring the Internet: Using the contraversial.] (1999, December 16). *Frankfurter Allgemeine Zeitung.*

*Net-Business.* (2000, June 24), p. 70. Source: http://www.ivw.de

Rammert, W. (1996). Kultureller Wandel im Alltag und neue Informationstechniken [Cultural change in daily life and new information technologies]. In J. Tauss, J. Kollbeck, & J. Mönikes (Eds.), *Deutschlands Weg in die Informationsgesellschaft [Germany's Way into the Information Society]* (p. 277). Paris: Baden-Baden.

T-online wählt ZDF als Nachrichtenpartner [T-online selects ZDF as newspartner]. (2001, March 21). *Financial Times Deutschland.*

van der Meulen, L. (1999, November 30–December 1). Speech about the regulations and content of the Internet, Summit of media regulators, Paris. http://www.cvdm.nl

# Policy

# 7

# Regulatory Concerns

Robert Pepper

*Federal Communications Commission*

The domestic consumer content industries in the United States include broadcasting, film, radio, TV, cable TV, computer application software, video games, Internet online content, newspapers, books, and so on. All of these account for almost one third of a trillion dollars per year (Fig. 7.1). In addition, U.S. domestic transmission network revenue amounted to one third of a trillion dollars per year. In addition, U.S. domestic transmission network revenue also amounted to one third of a trillion dollars in 1999 (Fig. 7.2). Thus, what is at stake is no less than the potential restructuring of two thirds of a trillion dollars per year in revenues, and the communications industries as they are known today.

New entrant competitors are beginning to enter the U.S. market, investing capital and building their own backbone networks. Investment in new U.S. networks and infrastructure doubled in real dollars over the 4 years beginning just before passage of the 1996 act to $60 billion per year, and ending with the downturn in dot-coms and telecommunications (see Fig. 7.3). The most dramatic growth has been in the new entrants in wireless communication. The incumbent local exchange carriers' capital investments in constant dollars were flat or slightly declining until they faced competition, at which point they increased infrastructure investment by about 15%.

There has been a competitive deployment of broadband among digital subscriber loops (DSL) and cable modem service (Fig. 7.4). But how far along is the deployment cycle? DSL is a technology from the 1980s. Cable modems were talked about in the late 1980s and early 1990s. The first commercial cable modem was deployed in 1995 and the first commercial

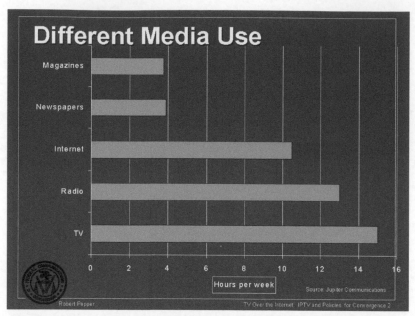

FIG. 7.1.　Different media use. Data from Jupiter Communications.

FIG. 7.2.　Demand for digital content: Consumer market for bits. Data from Veronis Suhler, *Communications Industry Forecast* (Nov. 1999) and 1999 U.S. Statistical Abstracts. Computer Software is a 1998 estimate by the Software & Information Industry Association.

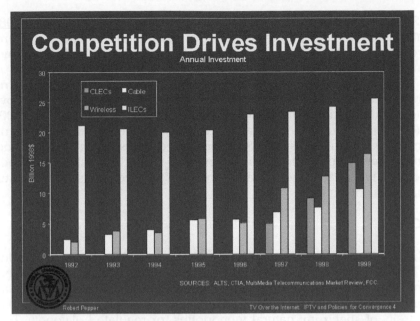

FIG. 7.3. Competition drives investment. Data from ALTS, CTIA, MultiMedia Telecommunications Market Review, FCC.

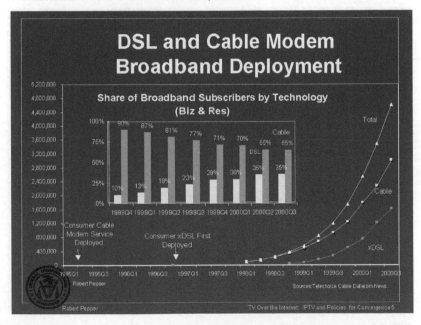

FIG. 7.4. DSL and cable modem broadband deployment. Data from Telechoice Cable Datacom News.

DSL in 1996. The inflection point was in 1998. At the beginning of 1998, there were about 50,000 cable modems. By the end of that year, there were about 500,000. The next year, 1999, began with 35,000 DSL lines and ended with 550,000 lines (see Fig. 7.5).

It is clear that a competitive dynamic between the local exchange carriers and cable companies developed that has driven competitive deployment and adoption of these two broadband services. At the end of 2000, there were approximately five million high speed Internet access consumers with cable modems and DSL in the United States. Things appear to be on the steep slope of the classic adoption "S-curve." This is important because it will enable the interactive market to develop, assuming other things do not get in the way.

## POLICY

The regulatory framework in the United States, and the rest of the world, is based on a set of old rules and assumptions. The old rules were based on distinct industry and regulatory structures. The old rules and regulatory structure were based on noncompetitive models. They assumed a scarcity of spectrum, natural monopolies, or regulated oligopoly. The

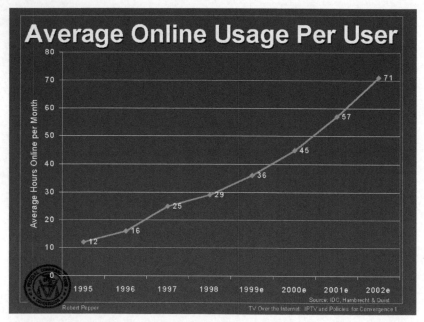

FIG. 7.5.    Average online usage per user. Data from IDC, Hambrecht & Quist.

government protected the incumbents, thus insuring scarcity. The thought was that it was necessary to regulate a natural monopoly. Over the last decade, it became clear that it was an unnatural monopoly, created to a large extent by government decisions to protect incumbents in order to stimulate investment. In the name of protecting consumers, the government created barriers to entry and decided winners and losers. Licenses were given to people who promised all kinds of wonderful things, and they were never kept to their promises. In fact, the result was what some have referred to as *regulatory capitalism,* where one invests in lawyers and lobbyists instead of technology and services. Those old rules are being challenged.

Digitization is driving the change. It creates conditions for competition, and reduces entry costs. Innovative services and products at much lower costs have been introduced. Businesses today have the flexibility to combine different products and services. CNN online and the *New York Times* Web site are new media combining text, pictures, streaming audio, and streaming video. All of the communications regimes are going digital. The industry boundaries are beginning to blur and digitization makes the old regulation of industry structures obsolete. The neat little silos used in the past are no longer applicable going forward.

Digitization also destroys compartmentalization. A bit does not know if it is a broadcast bit, a cable TV bit, a telephone bit, or a computer networking bit. The bit can be transmitted seamlessly; it is neutral to the medium of transmission. It can be stored, processed, manipulated, and forwarded. Intelligence can be everywhere in the network and at the edges. It is a false debate between smart networks and dumb networks. There will be both, with intelligence in the networks and at the edges. How all that sorts out will depend on the market and not regulation.

These new realities are confronting the old rules and structures. The old rules that were based on distinct industries and structures were written at a time when the nature of the conduit determined the content. The content came with the conduit. With digitization, these are no longer necessarily linked. This leads to interesting developments.

The traditional regulatory approaches that kept communications industries within traditional boundaries will not work anymore (Fig. 7.6). This is not just a matter of regulation, but of changing business models. This is difficult because change is especially difficult for incumbents, and in the telephone world some incumbents are 120 years old.

In the traditional world of telephony, information flow was always one-to-one. The capacity constraint was essentially the switch. The network was designed on the assumption that not everyone wants to call at the same time. When they do, the result is busy networks. In broadcasting, the architecture is one-to-many and the capacity constraint is the radio

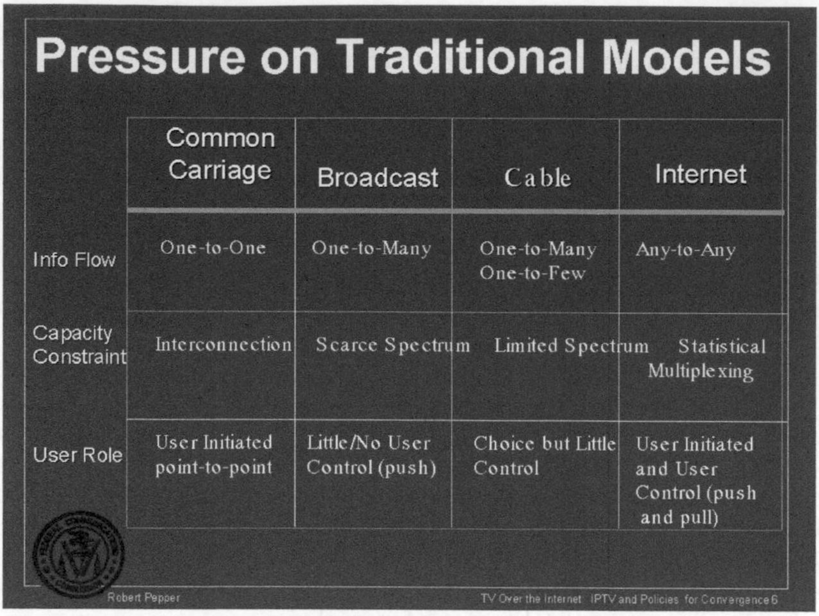

FIG. 7.6.   Pressure on traditional models.

spectrum. And, there is essentially no user control. TV and radio were the first effective push technologies. This architecture is extremely efficient when many people are watching the same thing, but it is constraining from the user's perspective. The nature of information flow on the Internet is many-to-many. The constraint is bandwidth, or the transmission speed of the connection. And the use is user initiated and controlled, push and pull.

There are a number of important questions facing IPTV. IPTV may refer to Internet protocol TV, interactive personal TV, or intelligent personal TV. It can mean different things. Actually, it is all of these because the market will determine how it develops.

Will Internet TV be evaluated by looking forward or backward? Is it necessary to look forward in terms of new market policy and consumer models or should the old models be allowed to drive what will happen in the future?

Intellectual property rights (IPR) are a major question. (See Carter, chap. 9, and Einhorn, chap. 10.) Will they be used as an enabler or a barrier? Who controls the content, the gateway or the customer? Although intellectual property rights holders need to be compensated for their intellectual property, 19th- or early 20th-century models for intellectual property rights protection and compensation may not work going forward. IPR may become one of the biggest barriers to the development or deployment of Internet TV whether it is Internet over TV or TV over the Internet. Copy protection issues and compensation are key when considering linking content and conduit. Intellectual property right protection issues are holding back the development of new media. People will need to figure out different ways to get paid. Fortunately, the willingness to pay is exceeding anything once thought possible. Eighty percent of Americans pay for their TV today in the form of cable or satellite. Because the willingness to pay is so high, it should be possible to develop a new IPR model to make it an enabler of the new Internet-based service.

## POLICY DOES NOT EQUAL REGULATION

There are many policy issues related to TV over the Internet. For truly interactive TV with purchasing on the web, how should privacy and information use be handled? In the cable TV world, there are very specific limitations on what the cable operator can do with subscriber identifiable information—basically nothing without subscriber permission. In the telephone world, there are limits on customer proprietary network information. Can there be comparable protections for TV over the Internet?

Another concern is how to protect children, not just from pornography but also from terrorism, gambling, or the sale of alcohol over the Internet. There also are issues concerning electronic commerce transactions over TV, over the Internet, not to mention taxation, tariffs, uniform commercial codes, and trust. These t-commerce issues and their resolution are really no different than the general e-commerce questions. However, there are people that want to impose general media rules on top of the Internet. That would be extremely counterproductive.

## INTERNET TV COMPETITION REGULATION AND POLICY

The question is, why is legacy media regulation necessary on the Internet when there is no scarcity? This goes to the question of incumbents creating scarcity and the government maintaining it in order to justify regulation to protect consumers because there was scarcity. Rules were used as a shield against competition. This is not new. Broadcasters argued for decades that cable TV would destroy broadcast TV and the consumers would be worse off. So broadcast TV was protected from cable for 30 years. Then the cable industry argued that satellite TV would harm cable TV. Each industry argues that competition is good just as long as they are entering someone else's market.

The Internet is not a traditional electronic mass medium. Attention must be paid to the incumbents arguing for a "level playing field" to slow competition. Putting those old media content rules on the Internet would be counterproductive.

The decision was made not to regulate the Internet by not imposing legacy telecom regulation, calling this "unregulation." Several lessons have emerged from this policy. First, there are real benefits from not imposing old rules designed for monopolies on new services and entrants. Second, when the new services compete with the legacy services, regulatory parity can be achieved by deregulating the incumbents rather than regulating the new entrants. Third, there is still a need to worry about bottlenecks and anticompetitive behavior, especially from former incumbent monopolies.

In general, however, regulators should be promoting rather than prohibiting. The issue at hand is not about traditional media regulation, but rather about enabling entirely new platforms for new, innovative, and competitive services that consumers will value. Policymakers will need to take the same forward-looking perspective as technologists and investors.

# 8

# The Challenges of Standardization: Toward the Next Generation Internet

Christopher T. Marsden
*Re: Think and The Phoenix Center*

The mass distribution of video programming over IP networks promises a richer experience for viewers, with widely predicted increases in interactivity, choice, personalization, and the ability to micro pay for a la carte programming.[1] Whereas broadcasting was licensed, controlled, and tightly regulated by national governments (or even owned as a monopoly service), video-over-IP will be delivered by international market mechanisms with both relatively minimal direct legal restraint and little direct government strategic intervention. Standardizing video delivery to produce network economies of scale and scope will require international corporate coordination between the converging industries of broadcasting and video production, wired and wireless telecommunications, and computer hard- and software derived data communications. In this economic analysis of law, I consider the distribution of existing television broadcasting archive over IP-based networks. While new production can be designed for IP networks in technological, economic, and legal terms, I postulate that it is access to the mass of video archive which will create the critical mass of online programming that drives the "video Internet." My focus is on the development of legal regimes based on market mechanisms, which will lead into the online exploitation of broadcast video rights. Although my perspective is predominantly European,

---

[1] See Waterman (2001). The regulatory implications in the U.S. are examined in Compaine (2001).

the markets are developing globally, and U.S. and Canadian law and corporate strategy are analyzed where appropriate. The overwhelming conclusion is that the Internet's engineering development is driven by the security, competition, quality, and reliability imperatives in monetizing broadband data, of which video is the paradigm I adopt. This development is achieved through international standardization by industry bodies supported by governments, and is emerging in creation of quality of service (QoS) in the local loop: the "final mile" to the consumer over which infrastructure and IPR owners exert control. This can only be achieved over broadband networks (see Shelanski, 1999), which requires investment in upgrading backbone (the "middle mile"), local access, and home access infrastructures. The investment required creates local monopoly and duopoly (the "last mile" issue), typically of fixed wireline access by cable modem and Digital Subscriber Line copper wire,[2] though other technologies exist to offer broadband wireless access (overcoming the "last metre" problem via 3G mobile and "4G" wireless LANs). Broadband networks are driven by the use of services that will monetize the bandwidth available. Following a summary of the state of video-over-IP legal, policy, and market developments required in sections 1 and 2 of this chapter (see also Marsden, 2001b), I examine in turn:

- In Section 3, the recent state of broadband market and policy development.
- In Section 4, TV intellectual property rights (IPRs) in the online environment.
- In Sections 5 and 6, I conclude that rights holders in infrastructure and Internet Property Rights (IPRs) will drive the development of a secure broadband local loop for delivery of IP video, signaling at least a temporary end to the Internet's founding architectural principle of "end-to-end."

The development of markets in real property (local loop and radio spectrum) and video IPRs has been severely hampered by the failure to delimit and efficiently transfer property rights. It is not an exaggeration to state that the development of the concept of property rights, together with a consistent and measured examination of the public interest in regulating and assigning those rights, are the primary challenges for both governments and market actors. This is beyond even the extraordinary pace of technological innovation that is creating the space within which those rights will be exercised.[3] North (1990) and Williamson (1975) have demonstrated that

---

[2]See variously Carter Donahue (2001), Eisner Gillett and Lehr (2000), and Faulhaber (2001), for definition and the open access debate. In this contribution I briefly address the open access debate in fixed line, concentrating instead on wireless infrastructures.

[3]For the regulatory and business challenges, see Figueiredo and Spiller (2000).

property rights are the basis for transferable wealth and therefore economic development.[4] The latter, following Coase (1937), has shown that where transaction costs in property rights are sub-optimal, corporations will be formed, internalizing those rights within organizations. Failing property rights transfer, economies are reduced to barter, in which roughly equally valued goods and services are exchanged without monetization. This paradigm, that without property rights being efficiently assigned, monetization of transfers is inefficient where possible, and replaced by barter, is the situation pertaining in much of the broadband media market. This applies to traditional broadcasters and Hollywood studios, but also to the music industry, and to broadband infrastructure owners. Monetizing this "barter" economy will require rapid evolution from the current IP infrastructure, as well as from the traditional broadcast model.

## 1. EVOLUTION OF THE INTERNET: BEYOND END-TO-END

End-to-end was the guiding principle in founding the Internet (Saltzer, Reed, & Clark, 1984). Kahn and Cerf (1999) embraced an all-encompassing definition of "the Internet as a global information system, and included in the definition, is not only the underlying communications technology, but also higher-level protocols and end-user applications, the associated data structures and the means by which the information may be processed, manifested, or otherwise used."[5]

It is a packet-switching network, with no dedicated channels. It delivers all the packets sent onto one end of the network to the other end. It makes no distinction between video and other data signals. Like trucks on the road, the packets mix with all other traffic before meeting the other trucks in the fleet at the destination. The Internet acts as a "dumb" network delivering to an intelligent box, the Personal Computer (or other micro-processing device), which decodes and orders the packets to make an intelligible message from the data packets. Fundamentally, the Internet is the 1960s ARPANet, in its present Internet Protocol Version 4 (IPv4) form. IPv4 is the basic standard of the any-to-any environment that "accidentally" became the global TCP/IP standard for data, voice, graphics, and audio. The idea was that the system would not discriminate between packets of data or users: Anything sent from one end would reach the other. It relied on the intelligence in the system being distributed in PC terminals, where data packets would be decoded. As the size and complexity

---

[4]For an insightful application to telecoms regulation, see the public choice assessment of the U.S. constitution's protection of private property (Cherry & Wildman, 2000). An excellent property-rights-based treatment of regulatory arbitrage in Internet governance is that of Burk (1999). An early path-breaking treatment of comparative telecoms regimes is by Levy and Spiller (1994).

[5]For the history of the Internet, see Leiner et al. (n.d.).

of data packets was relatively similar from one user to the next in the early stages of email and text pages, and could be easily reassembled by PCs, the end-to-end principle became enshrined as the founding principle.

The increasing bandwidth of the modern Internet, allied to broadband connections and more powerful PCs, has permitted much greater diversity in data, including graphics, audio, and video. This increases the complexities of traffic management, with rich media packets traveling at the same pace as "spam" junk mail. The next-generation Internet, IPv6, is expected to permit prioritization of time-sensitive and higher revenue traffic, introducing hierarchy into the system. The fear of the early "netizens" is that the innovation which the information-sharing, non-encrypted, non-hierarchical, "free" Internet will be undermined. As Lemley and Lessig (2000) wrote in opposing the vertical integration of telcos with cable TV firms, they fear "the end of end-to-end" (see also Berkeley Law and Technology, 2001).

The rough and ready non-hierarchical protocol that served defense, academic, and "geek" users until the mass adoption of the Internet in 1995, will be fundamentally altered by the new IPv6 that is being developed, challenging the any-to-any nature of the Internet ("Upgrading the Internet," 2001). Both Vint Cerf and Dave Clarke, pioneers of the original Internet, see an inevitable evolution to priority on user-pays in a "rich media" Internet, where Application Service Providers (ASPs) and streaming media delivered over Content Delivery Networks (CDNs) will occupy much of the data traffic delivered over TCP/IP networks. Enron Broadband Services predicted the data traffic for 2005 as shown in Table 8.1 (see also Morel [2001]). This demonstrates that streaming media will be a critical, but by no means dominant, data type on the next-generation Internet. It

TABLE 8.1
*Global Data Traffic on TCP-IP Networks 2005 (2000)*

| Data Type | Percentage of Traffic 2005 (2000) |
|---|---|
| Machine-to-Machine (e.g., file back-up; remote security) | 45% (37%) |
| Peer-to-Peer (e.g., file and application transfer on request) | 24% |
| Streaming Media (audio/video) | 21% |
| Web pages | 10% |

does show, however, that Web pages will be a far less dominant artifact of the next-generation Internet. To deliver this future video and shared ASP will require greater security and prioritization of data on the Internet. If you pay more at the mini-tollbooths that will monitor and check data packets, you will be safer, faster, and better able to plan your journey, business, and life. That will introduce more control, which rightly worries Internet purists. It holds the hope that the one-to-many broadcast channel will be supplemented by the delivery of video over the Internet.

Predictions of streaming media revenues are less precise, as Fig. 8.1 demonstrates. Note that the vast majority of revenues in 2000 were audio, not video. Yankee Group predicts broadband PC penetration will reach one third of all Internet-connected households, and 25% of U.S. households, by 2004 (see Fig. 8.2). Even if this figure is considered high, it demonstrates the development of a mass broadband Internet market. European communications policy had no end-to-end tradition, with government control and censorship of communications, and a less individualist notion of freedom of information.[6] End-to-end via the Internet raised

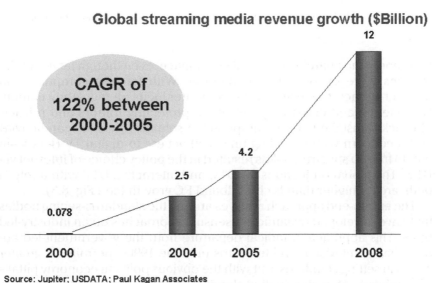

**Global streaming media revenue growth ($Billion)**

12

CAGR of
122% between
2000-2005

4.2

2.5

0.078

2000        2004    2005            2008

Source: Jupiter; USDATA; Paul Kagan Associates

FIG. 8.1. Global streaming media revenues 2000–2008 ($b). Source: Jupiter; USDATA; Paul Kagan Associates.

[6]On the European regulatory legacy, see, for instance, Blackman and Nihoul (1998) and Marsden and Verhulst (1999).

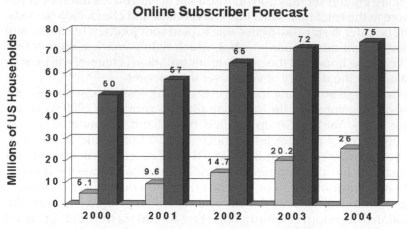

# US Market Maturing Fast

**Online Subscriber Forecast**

**Over a third of all online homes will subscribe to high speed access by 2004**

Source: The Yankee Group, 2001

FIG. 8.2.   U.S. broadband market maturing fast. Source: Yankee Group.

fears about the anarchic nature of cyberspace, even though far fewer Europeans were exposed to the Internet. With far less computer and Internet literacy, the result has been greater focus on regulating content (via filtering and classifying websites), protecting minors and privacy. The lack of installed PCs in European households (outside Scandinavia) and a concern with providing universal access to digital TV (DTV)—in part to free up spectrum—has resulted in the policy choice of Internet via DTV.[7] The global projections for DTV, and interactive DTV with a return path, are far higher than for broadband PC growth (see Fig. 8.3).

The end-to-end approach revolves around the standards-setting bodies that have developed dynamic consensual approaches on an industry-led basis. This approach, a radical departure from the government-led approaches of television and telecoms until the 1990s, permits far greater speed in setting standards, but with the obvious political economy pitfalls of dominant actors dominating standards bodies. Further, the resources necessary to influence the plethora of standards in converging media, telecoms, and IP environments on a global basis are available to only a few

---

[7]This creates huge problems of scarcity and access. In a voluminous literature, see, for instance, Cave and Cowie (1998), Cowie and Marsden (1999), and Flynn (2001). For updated analysis, see Marsden and Ariñe (2003).

# Global DTV Homes (m)

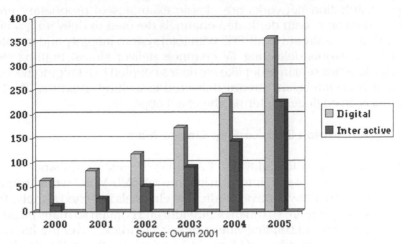

FIG. 8.3.   Global DTV homes (m). Source: Ovum 2001

large multinational enterprises.[8] While organizations such as the IEEE (Institute of Electrical and Electronic Engineers) and IETF (Internet Engineering Taskforce) claim that individuals leave their corporate identity outside the negotiation of standards,[9] the reality is far different.[10]

## 2. THE START OF VIDEO-OVER-IP AS THE BEGINNING OF THE END OF BROADCASTING?

The choice of device is critical to "end-to-end." TV is one of the least end-to-end of communications technologies: It has no interactivity (no "return path"), so cannot be end-to-end unless supplemented by telephone connection.[11] Further, it is a "dumb box" with an intelligent stream:

---

[8]For problems in industry-led standard setting, see Lessig (1996b), Marsden (2001a), and Kahin and Abbate (1995).

[9]Excellent reviews of the literature are contained in Besen and Saloner (1989) and Noam (1989). See further David and Shurmer (1996) and McGowan (2000). On the IETF and IEEE specifically, see Gould (2000).

[10]On the possibilities of open standard setting, see Bar et al. (2000).

[11]This creates massive legal definitional problems; see Marsden and Verhulst (1999) and McGonagle (2001).

The technology delivers perfect pictures by allowing almost no data other than TV pictures over the network. This is the opposite of end-to-end. Television distribution networks are classic examples of proprietary engineered networks, with dedicated channels devoted to delivering a high bandwidth data stream in MPEG2 with total system integrity and no interference. Networks delivering TV channels deliver almost nothing else; data packets are sent in order like carriages coupled to a locomotive on a track. It is an intelligent stream delivered to a dumb box: the TV. The Internet is, in engineering terms, the exact opposite.

## Analogue Television Broadcasting: One-to-Many

It has always been thought that TV signals (including enhanced DTV, which puts a layer of xML interactivity onto a 3-6Mb/Mbls) would require the equivalent of railway travel: dedicated channels for video signals. This is because they require so much bandwidth and contain so many packets. Early experiments in sending video packets by "truck," over the Internet, have been hampered by accidents at junctions on the network. When packets hop from one network onto another, they can be delayed or even lost at the junction; hence buffering and frame loss in the final picture delivered to the viewer.

Delivering TV by dedicated networks has its disadvantages: Choice is limited, flexibility is lost. Market researchers and advertising agencies have proven beyond all reasonable doubt that consumers want individual choice in programming (Arbitron/Coleman Research, 2000). Broadcast technology can't cope with choice; it is a network built for mass transit by trains, not individual trucks. Broadcasters have persuaded governments to introduce DTV. These "trains" are bigger (interactive TV can occupy up to 8Mb/s) but much more efficiently timetabled (multiplexing allows up to 6 channels to occupy the spectrum of a single analogue channel). They are more expensive for the consumer—government compulsory licence fees have increased overall in Europe to fund DTV—and in the case of pay-TV, much more expensive, with subscriptions at up to four times analogue licence fees.[12]

## Starting Video-to-Many: Developing the Broadband Internet

For increasingly complex data such as voice and audio, TCP/IP packet switching is rapidly taking ground from switched circuit networks. For video, which has been considered too huge and delicate—like transporting bulk chemicals—it is both clearing the last mile, via tolls, and shrinking

---

[12]For a critique of the regulatory capture involved in such a move, see Marsden (2001b). On the theory of regulatory capture, see Moe Terry (1997). See also Marsden and Ariñe (2003).

the packets to make transport easier, which will let those of us who want to get off the train and drive ourselves to do so. Moreover, the trucks will need to be tracked as they move to ensure they don't get lost, taxed on their progress so that government takes its share, and reassembled seamlessly at their destination depot. It will also help if smaller loads can be delivered flexibly to individual customers: interactive services and advertising as well as on-demand programming. That needs the same network, but more intelligence at the depot. Instead of the terrible cart paths and narrow roads on which academics built the original Internet, new autobahns have recently been constructed by fibre-optic providers such as KPN Qwest and Level3. At the junctions, Content Delivery Network (CDN) companies such as Akamai have built new toll lanes, taking traffic quickly into the center of town (from where you make your way slowly down the last mile). Traffic is still slower than on the trains, but with huge advantages in freedom of choice. So, how to squeeze packets onto the Internet?

1. Widen the highways.
2. Compress the packets.
3. Track the packets automatically as they travel.
4. Deliver to households as efficiently if they were on motorways.
5. Decompress the packets between the depot and the viewers' houses, and reassemble in perfect coordination at the destination.
6. Persuade the viewer to pay at least as much to receive the packet by IP rather than broadcast networks, by offering greater flexibility.

Finally, the vested interests on the trains, switched networks, will need to be persuaded to accept robust competition. That also means persuading government, via content filtering and anti-piracy, to stop moral hazards being used to prevent the market from developing. Table 8.2 presents ways in which networks determine interactivity and efficiency.

Before examining in section 4 the IPR issues in adapting current packets of video to the Internet, I first examine in some depth the wired and wireless attempts to bridge the "middle mile," "last mile," and "last metre" (or yard).

## 3. BROADBAND BOTTLENECKS: THE "MIDDLE MILE," "LAST MILE," AND "LAST METRE"

The lack of legal certainty in assigning property rights is restricting the growth of a broadband Internet, and leading to a localized, Balkanized "walled garden" private network approach: back to the future. In such a fragmented future, should it continue, the issue of open access to those private networks will become critical. The global information infrastructure is becoming increasingly regional, national, and local, a process

TABLE 8.2

*How Networks Determine Interactivity and Efficiency*

| Network Properties | Broadcast | Video Over IP | Infrastructure Improvements Needed |
|---|---|---|---|
| Bandwidth | High 3–6Mb/s | ADSL: 512Kb/s+ POTS: 56/33k | "Last Mile" DSL, Cable, 3G 500–10,000kb/s |
| Two-way Interactivity? | Very limited in MPEG2 | ADSL: High Satellite: Low | High Bandwidth Return Path |
| Packet Size | Huge | Low And Reducing | MPEG4 Standardized 2001 |
| Monitoring | NA—closed network | Low But Increasing | Digital Rights Management |
| Reassembly | NA | Good And Improving | Improved IPv6 Internet |
| Delivery | Consistent | Poor But Improving | "Middle Mile" Hops Between CDNs |
| Cost | Low | High But Decreasing | Virtuous Scale Economies Circle |

which streaming video will accelerate due to the huge (probably insuperable) technical and legal challenges it represents to any global broadband solution. Without a more legally certain international allocation of property rights, the old national legal restrictions will continue to apply to profitable mainstream operators, with the global public Internet a source of piracy, romance, and buccaneering on the high seas beyond the reach of national legal certainties. Delivering a regulatory and market proposition to make the highways affordable to businesses and then to consumers (possibly all consumers, eventually) is a huge challenge, especially in the final mile and final metre delivery to households. Government spectrum auctions have thus far proved an inefficient digital alternative to the analog of building a national highway system, but the technology is expanding choice so fast that the market may deliver with minimal government interference beyond unbundling the local loop.[13]

---

[13]As an example of market innovation, see, for instance, the IEEE 802.11a wireless local area network (WLAN) standard that can deliver 54Mb/s from a base station which could cost less (much less) than $1,000. See also Croxford and Marsden (2001).

## "Middle Mile" Bottlenecks

The solution to the global Internet dilemma is in two parts: delivering content efficiently, in terms of speed and cost, and securing it from unauthorized use. The CDN solution is well known, using broadband backbone speeds and local hosting on proprietary networks—effectively avoiding the Internet wherever possible. It can be argued that "thinking local" works better using local closed telecom networks in each national geography to deliver content close to the user. By choosing the public IP "cloud" as a global solution, CDNs such as Akamai have to overcome the Internet's latency "middle mile" problem. The Internet is a network of networks; content delivered over the WWW has to "hop" from network to network, slowing it down at each hop. Both Digital Island, with its international backbone minimizing hops, and Akamai, with servers placed in each national geography, and often multiple networks within that geography, try to combat the "middle mile." They do this by cutting down the number of hops, ideally to two, onto their proprietary networks, and then onto the local partner's network to the home or business end-user. Unfortunately, there is always a "hop" or two too many to deliver at maximum efficiency. Those "hops" mean delays. Market surveys (for instance, Yankee Group) reveal that broadband has transformed the consumer Internet browsing experience, cutting out dial-up delays. While CDN solutions speed up delivery, they nevertheless require patience in broadband consumers whose raison d'etre on the broadband Internet is impatience. This wouldn't matter if there were no other way to access video as fast as Digital Island and Akamai. Unfortunately, there is; cable and satellite DTV.

Pay-TV avoids delays—technically termed *latency*—by using a satellite or proprietary fiber-optic cable to directly feed a local "head-end" or consumer dish, totally circumventing the "hops" over the WWW. Of course, this solution presupposes that video packets access the local network, where the gatekeepers—including Time Warner cable and your local telephone company—have no general economic incentive to carry these huge unwieldy and often revenue-losing packets to the end-user. As a result of the unsustainability of this revenue model in the investment climate of the summer of 2001, the leading global CDNs were taken over by backbone operators: Digital Island by Cable & Wireless, iBeam by Williams Broadband. To incentivize local loop gatekeepers, IPR owners have increasingly decided to cut deals direct for locally cached content rather than hopping over the Internet on CDNs. Consider first government incentives to encourage broadband local loop investment.

### "Last Mile" Bottlenecks

The gatekeepers face two massive property rights challenges, which have become especially profound in the European market: third generation

(3G) wireless spectrum rights and local loop unbundling (LLU). Globalization of the telecoms industry creates tensions between national regulators.[14] This is most obvious in policing of the electromagnetic spectrum, where a scarce resource must be shared and new services planned so that technical interference is minimized. However, national regulatory differentials have caused chaos in the auction of 3G licences for mobile telephony.[15] The failure to coordinate a common standard for the European 3G auctions is one of the twin tragedies of member-state regulation of telecoms in 2000. The other is the continued failure to develop alternative local loop broadband services, by divesting the telco of its cable TV division in advance of the 1998 local loop liberalization to permit upgrading of services from analogue to digital.[16] A partial answer has been to "unbundle the local loop" by co-locating rival operators' switching equipment in local telephone exchanges, permitting them to use the higher bandwidth element in the copper wire lines for Digital Subscriber Line (DSL) services. With the combination of competitive infrastructure being least advanced and regulators most advanced, LLU has been a partial success.[17] This "managed competition" to the incumbent creates huge regulatory distortions of market valuations.

Competition is seen to be emerging in broadband via these two routes. The regulator is opening access to the assets of the telco—wedded to "midband" 128Kb/s Integrated Service Digital Networks (ISDN) previously. The telco rationally fears that DSL would cannibalize ISDN revenues. Whereas previously, mobile was held to be the most potentially profitable market followed by local loop, the effect of auctioning 3G and regulating broadband local loops has contributed to the decision by some European telcos to divest their mobile divisions. The experience appears to have made fixed returns fluctuate wildly in the sector. Regulators need to ensure that basic network integrity survives, that 3G networks are built on time, and that rival DSL operators do not leave the

---

[14]For critical commentary, see, for instance, Naftel and Spiwak (2001), Laffont and Tirole (2000), and Marsden (2000). A comprehensive and complete analysis is offered by Larouche (1998).

[15]This was caused by the high cost of the UK and German auctions, the low cost of the French and Swedish "beauty contest" auctions, and the "middle way" in Holland and Spain. Regulatory chairmen Martin Kurth of German RegTP and Jens Arnbak of Dutch OPTA have described how mobile auctions had caused the cost of capital to rise for national telcos.

[16]The extent to which liberalization has involved hugely increased regulation is demonstrated by the 2001 legislative program (Commission of the European Communities, 1999).

[17]A useful measure of local loop competition is supplied by the Competition Scorecard maintained by the European Competitive Telecommunications Association (ECTA) at http://www.ecta.org

market, driven out by regulatory uncertainty in assigning a new property rights settlement. The property rights shambles proves the poverty of regulatory zeal, with spectacularly high mobile auction prices, intransigent telco (and cable) management, and national regulation of LLU, which have delayed broadband roll-out.

The possibilities of alternative wireless access are considered in the following section, to illustrate the problems that can arise in creating an "open" alternative standard for IP transmission to the local loop.

### "Last Metre" Bottlenecks: Wireless Local Area Networks (WLANS)

Providing in-building wireless broadband networks is now feasible for both corporate and consumer premises, removing the need for multiple cables, and fixed line Internet access. Until recently, chipsets were unable to practically reassemble wireless broadband multiplexed signals, but Moore's law has overcome that microprocessor problem such that Calgary-based WiLAN predicts 155Mb/s download speeds by 2003. These capacities are far superior to 3G mobile telephony, which is expected to achieve only 2Mb/s from each base station. It is suggested that integrating 3G mobile with WLAN can help to achieve localized broadband in populous areas, with lower but still always-on packet-switched capability between these local "hotspots" of broadband. However, the engineering and standardization challenge of WLAN is considerable. As with 3G, it is complicated by rival U.S., European, and Japanese standards. WLANs operate in the 2.4GHz and 5.4-5.7GHz GHz bands, in the Industrial Scientific Medical frequencies, which are unregulated. Consequently, reception in potentially crowded and "noisy" (full of interference) spectrum requires sophisticated and standardized devices. The process by which devices are standardized differs according to market, but the major standards setting institutions are the IEEE for the United States, MMAC for Japan, and ETSI for the European Union (see Table 8.3).

*United States: IEEE and 802.11.* Standardization of WLAN in the United States is carried out by the IEEE, an engineering body that provides a self-regulatory solution. The standards family is 802.11, and the first-generation standard is 802.11b, from which upgrade to second-generation 802.11a will take place. The 802.11 working group's voluntary standard is certified by the Federal Communications Commission (FCC), the federal agency responsible for all U.S. communications. Previously committed only to domestic use DSSS-compliant standards in the 2.4GHz band, the FCC set an important precedent on May 11, 2001, by admitting that its rules were out of date and decided, subject to consultation, to accept WiLAN's W-OFDM standard as well (FCC, 2001). IEEE standard-setting sets precedents that the FCC tends to follow. On security and QoS issues required to

**TABLE 8.3**
*Standards for WLANs*

|                     | European              | United States          |
| ------------------- | --------------------- | ---------------------- |
| Standards Regulator | ERO - ETSI            | FCC - IEEE             |
| 2.4GHz at 11/22Mb/s | Bluetooth; HiperLAN1  | 802.11b–WiFi; Home RF  |
| 5.4GHz at 55Mb/s    | HiperLAN2             | 802.11a                |

monetize WLAN services, IEEE 802.11 Task Group E is now working on security, range, interference issues, through the Media Access Control layer (MAC) for 802.11 platforms.[18] At the time of writing, the 2.4GHz band is unregulated, but crowded and "noisy" as a result. Services are expected to migrate to 5MHz band.

*HIPERLAN2: The European Answer to 802.11a.* HiperLAN2 is the European upgrade from the basic functionality of HIPERLAN1, a standard that was overtaken by commercial development of Bluetooth.[19] In part, this is due to Bluetooth's slowness to market. It is not as expandable as 802.11b; hence the HiperLAN2 upgrade option, which is intended to outperform 802.11a. HiperLAN2 is claimed to offer greater interoperability with 3G mobile networks, given its different MAC layer developed on the telco ATM technology, rather than the IP evolution of 802.11.[20] The European Radiocommunications Committee (www.ero.dk) has recommended that 802.11a devices not be permitted in member states until dynamic frequency selection (DFS) is enabled in PC cards, thus equaling QoS of HiperLAN2.[21] If national authorities hold strictly to this, that is likely to prevent 802.11a roll-out until 2003, diminishing the threat to dominant equipment vendors' 3G network build-out. Such a decision not to permit co-existence would have prevented the market's decision to adopt Bluetooth rather than HIPERLAN1, and ultimately 802.11b globally. Jippii,

---

[18]The result is likely to be upgraded capability for 802.11b, such that it offers a basic version of 802.11a capability, but with less range and lower security. The IEEE MAC specification applies equally to both 802.11b and the next-generation 802.11a.

[19]Windows XP supports only WiFi: see http://www.zdnet.com/enterprise/stories/wireless/0,11928,5080760,00.html

[20]Implementation of the HiperLAN2 standard, Annex 1P of ERC Recommendation 70-03, was on March 21, 2001, complete in six countries (Cyprus, Estonia, Finland, Iceland, Norway and the UK) and planned in most others.

[21]See http://www.vnunet.com/News/1117516

the most advanced European operator in 2.4GHz roaming service, can up-grade to either HIPERLAN2 or 802.11a. Unless the technical argument is overwhelming, co-existence of standards is always preferable, where the market can decide which offers better value to the consumer.

## Competitive Standard Setting

In the 3G standards battle, Grindley, Salant, and Waverman have empha-sized the use of new voluntary trade association bodies setting non-man-datory standards, which would suggest IEEE flexibility before ETSI certainty. While acknowledging the potential this creates for free riders and market-led innovation to overtake the standards process, they con-sider that these risks also encourage more rapid decision-making.[22] Volun-tary standard setting also permits co-existence of standards, and prevents a dormant standard being adopted, because the market judges rival stan-dards and will in all likelihood choose a winner. The issue of its engineer-ing integrity is relegated to a secondary consideration beside its ability to satisfy a timely market need. The 2.4GHz spectrum provides the ideal op-portunity to experiment with unregulated commercial spectrum, combin-ing as it does an existing unregulated resource with clear upgrade path to 5GHz, and the tradition of IP standards, where QoS and non-interference are the responsibility of manufacturers and operators acting in voluntary enlightened self-interest.

## 4. IPRS AND VIDEO-OVER-IP[23]

IPR has, if anything, an even less certain set of property rights than the "real" property of local loop, mobile, and WLAN networks. This chapter has shown that broadband networks make it possible to offer real video-on-demand (VOD). This development depends on releasing the IPRs in video properties for distribution over new media. There are two problems: The owners don't want to do it, and the rights don't exist.

The owners don't want to do it for reasons of bounded rationality. First, they are making supranormal returns already on their broadcast busi-

---

[22]As with 3G standard-setting, non-U.S. corporations fear that voting and techni-cal assessment procedures are biased towards "home town" players in the U.S., and U.S. corporations fear the same in the European standardization process. See Grindley, Salant, and Waverman (1999).

[23]I am grateful for interviews given in the course of research in winter 2000–2001 by many sources, most of whom maintain commercial confidentiality. I gained great theoretical and practical insight from discussions with Pamela Samuelson and Mark Lemley from Boalt Hall School of Law, at the University of California at

(continued on next page)

nesses, as most video rights holders are broadcasters granted monopoly, or at least severely rationed, licences. The long-term prospect of increasing revenues via VOD over broadband connections is outweighed by the short- and medium-term prospects of sustaining advertising and pay-per-view revenues in the rationed broadcast space. The prospect of more perfect competition in broadband is therefore not at all appetizing. Martin Tobias (2000) has stated that it is "Capitalism 101" that you must offer IPRs both protection and monetization; the Internet offers neither. The Internet must be made: (a) faster—by localized caching in Content Delivery Networks (CDN); (b) safer from IPR piracy—by digital rights management (DRM); and (c) potentially more profitable—by content syndicators who take audio, text, and video from hundreds of suppliers and supply to thousands of websites. IPR owners, notably broadcasters, see a much better future in using proprietary networks to distribute their video product.[24] Sandelson (2001) has demonstrated that there is no satisfactory allocation of IPRs for Internet distribution of video, where the TV rights already allocated are national in scope but Internet distribution requires global rights. The answer increasingly employed is to use the guaranteed service quality and enhanced security of the "walled garden" broadband service providers' network, to avoid the public Internet altogether. These "walled gardens" have a very satisfactory legal status; they are cable networks. The private network ensures integrity of rights, video delivery, and allocation of property. Unfortunately, in most European countries, except Sweden which has 10Mb/s to the kerb, streaming video to a full-screen TV in VHS quality is not possible in the "mid-band" bit rates available, 512-1768Kb/s, in consumer offers. Only truly private networks leasing high bandwidth at 2.3Mb/s, such as the UK Video Networks, can so far offer this walled garden service. It appears that migration to broadband Video-on-Demand requires a leap of faith by both telcos and broadcasters. The legal framework will ensure that this broadband VOD, when it arrives, will be more the

---

*(continued)*     Berkeley, and conferences held at Berkeley in March 2001 and the NYU Law School (2000) *A Free Information Ecology in a Digital Environment Conference* on April 2, 2000. In Europe, I am grateful to Bernt Hugenholtz of the University of Amsterdam Institute for Information Law (IviR) and attendees at a Council of Europe sponsored seminar in June 2001. See Hugenholtz (2000). The London law firms of Denton Wilde Sapte, Olswangs, Harbuckle and Lewis, Clifford Chance, were all invaluable in confirming the fragmentary nature of video rights. Interviewees from the BBC, Independent Television Commission, Bazalgette Productions, BTOpenworld, Aardman Productions, Loudeye, and Producers' Alliance for Cinema and Television were all invaluable. Especial thanks go to Pete Ward of Anonymous broadband consultants. All errors and opinions remain my own.

[24]Audio differs in that most radio stations are advanced effectively syndicator-aggregators of music files, producing little or no original material of worth. TCP/IP technology already permits music file theft on a grand scale, via peer-to-peer networks such as Gnutella.

AOL-style "walled garden" than true open access: private cable, not public Internet. Any solution that fails to acknowledge and cultivate the rights holders' strategy is attempting to rewrite the entire history of video sales, not to offer a value proposition.

This section is made up of three parts. In the first, the rights strategy of the video industry is explained. In the second, Internet-based offerings are examined, together with the disjuncture between their offer and the broadcasters' preferred environment. In the final part, I explain the broadband local loop solution and its "fit" with rights-holder expectations. In conclusion, the essential elements of any rights strategy are revisited. It will be seen that rights-holders expect the value created in the private controlled broadcast environment to be maintained in the broadband environment. It is concluded that only a closed private network can currently offer rights-holders the integrity to:

- Extend their brand on-line.
- Enhance the service they provide to their viewers.
- Monetize the value dormant in their archive.

### Video Industry Rights Strategy

The types of contracts that control video content are varied, but they revolve around one central factor. Before 1995, most content was not contracted for Internet or other forms of distribution. Primary broadcast rights for the national market were held by the broadcaster. Secondary rights, international broadcast, and distribution in forms other than broadcast, were individually negotiated. Tertiary rights to promotion, merchandising, and other forms of exploitation, were also a matter for negotiation. Some broadcasters became aware of multimedia at an early stage, and adopted the terminology "all media current or in future invented" to cover all forms of online distribution. An even more thorny ownership problem than primary, secondary, and tertiary rights to distribute on-line is third-party rights. A dramatic scene often involves a catalogue of third-party rights: two types of music (author and performer); producer; writer; actor. All these parties are represented by "collecting societies," a cartel formed to represent the individuals concerned. There is content which is off-limits, also self-identified by archives. This is typically archive with complex third-party rights (e.g., drama), and especially pre-1995 rights, where no platform is identified with broadband. The choice of distribution platform is critical.

The basic description of the TV rights legal framework demonstrates that every piece of content has a legacy of rights clearance. To reinvent the wheel is to seek global rights to compelling content with inadequate legal protection of property rights, or monetization to all parties of those rights: It is the Internet.

## Internet Distribution of Video

Internet streaming video claimed to "change everything." That is correct in that there is:

- No recognized geographical market.
- No rights holder revenue proposition except by cannibalizing existing revenues.
- No accepted industry standard solution which prevents piracy.
- No means of ensuring VHS quality streaming to consumers.
- Limited personalization and data mining for rights holders' properties.

Given the cry that "music wants to be free" of the Napster/MP3 generation, the answer from the recording music majors has been unequivocal: It will have its own credit card. Internet distribution of professional media products will be encrypted, secure, and monetized. Broadband IP networks have permitted distribution of digital recordings, and the downloading and file-sharing of Gnutella, Napster, and MP3 have created an environment in which the music majors have found themselves forced to distribute. Andy Grove stated in June 2001 that the media industries were at "their most critical inflexion point of all time ... they must decide the price point at which the majority of users will be honest" in paying for their products.

The video industry arm of these conglomerates has, unsurprisingly, adopted the same tactics as the music majors. Given the far greater technical complexity of video over audio, necessary to "capture the exponentially greater share of the individual's attention span," there is a short interval before the video industry reaches the "inflexion point" which university dormitories in U.S. college campuses forced on the audio industry in early 2000. Individuals have been prosecuted for cracking the DVD code off-line (*Universal Studios, Inc. v. Reimerdes*, 2000), and the gaming community is rapidly overtaking the video industry in creating effective codecs for secure video transfer. The standards community has gone far further in creating secure and high quality file transfer over IP networks for video. Rather than the relatively simple—and therefore easily cracked—MP3 format, the video industry has adopted increasingly high-end solutions. Streaming video increasingly adopts technologies based on MPEG4, with MPEG7 and MPEG21 emerging as the metadata standards that will create the "credit card" for individual content packets.

For IP-based streaming video syndicators and providers, QoS problems have thus far proven insurmountable. Three particularly merit-worthy attempts have been made to solve the QoS deadlock:

- Akamai is a partially distributed server architecture that aims to ensure higher QoS than the IP cloud without the investment available to provide a wholly private network. However good the server network, it cannot guarantee a VHS quality service. The engineering fact remains: The IP path is as fast as its slowest switch. If you do not control all switches (middle mile, last mile, and last metre), you cannot guarantee QoS.
- Loudeye aims to enhance the video experience by encoding and syndicating content—providing the credit card. The Loudeye Media Syndicator is a relatively sophisticated attempt to recreate the MPEG4/7 environment over the public IP network, providing a fairly high barrier for hackers. To eliminate IP theft of content, rights-holders know that they need to avoid the Internet altogether.
- Atom Films (merged with Shockwave) traded in content which is "designedly degraded." It is "optimized" for the Internet because it is poorer quality than VHS, and therefore viewable relatively easily at 300Kb/s, when that can be achieved. Their content consists of animation and short films, where creators accept the degraded product quality and viewing experience in the interest of branded global distribution. Other sites using degraded quality include adult, news (newsplayer.com), and rights-holders' promotional sites, where music and trailer promotional content is shown as teasers for the "main event" on TV, at the movies, or on VHS and DVD.

The Internet is thus being improved, security improved, and content "reduced to size." These all remain partial, hybrid answers to the basic conundrum: how to monetize archive over non-broadcast networks? The IP cloud cannot be the answer, as there is no recognized geographical market and its poor QoS risks cannibalizing existing revenues. Instead of adding value by monetizing content, it removes value by removing the key professional differentiator, editorial and production integrity.

### Broadband Local Loop

The complexity of the rights process lends itself to three main conclusions:

- Isolate that content in which rights are resolved for TV.
- Construct a distribution platform with similar legal characteristics to TV.
- Select a distribution method that creates as compelling an experience as TV.

These are the key legacy characteristics of video IPR rights:

- Video rights are assigned according to legacy agreements.

## TABLE 8.4

### Summarizing Content Owners' Dilemmas and Local Loop Solution

| Internet Rights Holder Dilemma | Example | Narrowband Answer | Loss of Market Value | Solution |
|---|---|---|---|---|
| International Internet rights quagmire. | Who owns the Olympics in Germany or Switzerland? | Clear all international IP rights, or none at all—loss of control over rights territories. | Failure to release full value from rights; cannot be windowed and leveraged; "one sale equals all" | Broadband local loop as the new rights paradigm:<br>• Closed proprietary solution using state-of-the-art DRM<br>• Permits local market sub-licensing. |
| Plethora of rights third parties | UK: Writers' Guild; BPI; Mechanical Rights Society; BECTU; Equity; PACT; Musicians' Union. | Use only pre-cleared or promotional clips produced for marketing purposes. | Free content sites pirating IPRs; proves value of experience but does not unlock value from archive. | Short format:<br>• Permits repurposing of existing content as "substitute" for trailers.<br>• Negotiation with key broadcasters confirms solution. |
| No IP rights pre-1995 | Assignment of rights completely omits on-demand network delivery. | Use post-1995 rights: no classic content; rights inflation for modern properties. | All classic archive lost to IP; over-valuation of non-compelling newly created content. | Assign clearances to IPR holders:<br>• Incentivized by PPV<br>• Short format<br>• International distribution<br>• Eliminate residuals |

*(continued on next page)*

| | | | | |
|---|---|---|---|---|
| QoS concerns prevents release of VHS and enhanced formats. | Film majors refuse to release sub-VHS buffered content; talent refuses to allow degraded delivery of product. | "Close to the edge" delivery using Akamai and others; MPEG4 permits greater compression. | Only low video grade content released: animation; pornography; audio; Shockwave. | • True edge delivery.<br>• Allow rights holders to trust and release best-mastered content.<br>• Virtuous circle of enhanced content and enhanced delivery. |
| Piracy concerns with public Internet. | DVD code cracked; MP3 solution for video now possible with DVD. | Watermarking (SDMI), DRM, standardization initiatives using BCDForum etc. | IPRs holders refuse to release content, editorial integrity offline – e.g. via DVD. | • Proprietary networks allow control of content delivery;<br>• Localized server delivery ensures security akin to broadcasters' own closed networks. |
| Lack of customer information prevents true eCRM value in exposure of rights holder property. | Advertiser dollars diverted from authenticated brand-building experience in broadcast to "anarchic," identity-theft prone delivery over public IP. | Value in rights hidden; existing "rich media" advertising offers fractional value of true rich media. | Stakeholders refuse to "cannibalize" existing revenue sources for low-grade alternative despite consumer demand. | • Complete customer information retrieval.<br>• Personalized advert delivery and personal content selection<br>• Broadband content value exceeds broadcast on per-viewer basis. |

Local loop works with this tradition—it buys based on traditional TV markets.

- Video rights are negotiated within distinct, generally national, territories.

Local loop is designed to assign national, and even local, territories.

- Video rights take no account of specific platforms.

Local loop is platform-neutral and a truly convergent solution, based on the most advanced screen for personalized entertainment as its first platform: the PC.

- Video demands high QoS and advanced anti-piracy protection.

Local loop offers VHS-equivalent streaming, with Digital Rights Management strategy.

Broadband local loop is not an Internet-based solution. It distributes via partner broadband networks an entirely private solution, with guaranteed QoS. With no public access, combined with advanced codecs developed for the video games industry and private network anti-piracy measures, broadband local loop ensures secure, reliable reception at VHS-quality. In some respects, it is better than VHS: It is hosted on the only digital screen in the household, the PC monitor; it is more secure than VHS tapes; it can be upgraded to DVD-quality over time. It can also ensure delivery to a single Point of Presence (POP), a local community, in the same way as cable television, permitting more discrete territories than satellite or terrestrial TV, or the Internet.

The choice of the PC as platform is also crucial. Broadband local loop can provide full-length programs over TV, directly competing with broadcasters' own classic archive channels. Home Choice (the Video Networks subsidiary) is competing with its suppliers on the same platform, "eating their lunch." In response, Hollywood studios have offered video-on-demand at prices that are non-competitive with Blockbuster, their preferred supplier, or Sky TV, the pay-TV operator. TV-on-demand cannot compete with the TV broadcasters and film studios' preferred distributors. Broadband local loop, by contrast, offers a method of monetizing selections from programs on a different platform to a richer, younger, more influential demographic, with the promise of far greater interaction, personalization, and e-commerce opportunities for rights holders. As the market gains consumer acceptance, users will create new submarkets based on genre to explore yet more of the archive and production capacities of broadcasters. This final point is critical: Local loop provides a new discrete revenue stream to rights holders. This provides both a promotional opportunity on the only proven e-commerce platform and a new discrete revenue window to rights holders, separate and complementary to existing broadcast and video sell-through windows. Local loop can granularize viewing to the individual clip level providing a further incentive: a level of market research and real-time market intelligence to rights holders, advertisers, and e-commerce partners never previously available. Local-loop rights strategy is a win-win game. The increased volume and quality of usage of local-loop broadband should help ensure that investments in broadband ISPs pay off, and further QoS, personalization, and content choice results, in a virtuous circle leading to the next generations of networks.

## 5. VIDEO-OVER-IP STRATEGY: COMPETITION AND COPYRIGHT

All innovative companies would logically prefer to monopolize their industry, while ensuring that upstream and downstream competition was sufficiently strong to create supply and demand efficiencies that would help to strengthen their hold on the most profitable link in the value chain. Achieving this goal has been critical to the success of Intel and Microsoft in the personal computer industry. Their respective domination of microprocessor chips and the Windows operating system has enabled them to secure huge margins on their core businesses. By sharing elements of the underlying code (but not the source code) with programmers and hardware manufacturers, they have ensured ruthless competition in PC manufacturing and software program development based on their platforms. They thus act as Wintel gatekeepers in the value chain, but also encourage innovation and competition in associated markets (see Lemley, 2000). Ensuring control of the gatekeeping function must not arouse the ire of the competition authorities. Intel has succeeded by largely confining itself to its core markets, and sharing code in a relatively non-discriminatory manner. Microsoft has entered downstream markets for applications running on its operating system, including the Internet browser market. What both achieved (though Microsoft's Supreme Court case outcome was uncertain at the time of writing) thus far is to convince competition authorities that the dynamism of their industries creates low barriers to entry and therefore that the lack of serious competition does not in itself indicate noncompetitive market conditions ("Guilty," 2001, *U.S. v. Microsoft Corp.,* 2000).[25] The entry of Linux and AMD to the operating system and microprocessor markets has helped to convince investigators that the potential for rapid erosion of the Wintel dominance exists.

Competition authorities appreciate that network markets combine this dynamism with plenty of opportunities for long-run dominance as consumers tend to rely on the standards of the dominant firm. As a result, especially in Europe, dominant communications actors have been significantly impeded in their search for dominance; Microsoft in its cable TV investments, AOL-Time Warner in its online music activities with Bertelsmann and EMI, Vivendi in its sale of pay-per-view movies of its new acquisition, Universal Studios. The new level of complexity in bottleneck analysis is the potential for perverse policy results arising from copyright and other IPRs. So long as markets can be isolated, copyright, government-sanctioned monopoly, is considered beneficial in creating innovation. Where various IPRs are bun-

---

[25]For commentary and a critique of Microsoft's defense, see, for instance, Lessig (1999a) and Liebowitz and Margolis (1999). For the European Union approach, see the recent decision by which Microsoft cable TV decoder investments were "neutralized" (Commission of the European Community, 2001).

dled together, in separate ownership, the creation of a bottleneck may be almost inevitable (Gifford & McGowan, 1999; Samuelson & Opsah,1999).[26] This was shown to be the case in all previous distribution media: the phonograph (record player), radio station, video recorder, audio cassette recorder. Peter Jaszi noted that:

> Section 111 of the 1976 Copyright Act cut the knot our courts had tied around cable television and unleashed a transformative force in the entertainment industry. Section 119 was introduced and extended in 1999 to provide a space for direct broadcast satellite technology. Compulsory licensing often has helped to open other promising channels for delivering content by breaking a decade's old standoff around performance rights and sound recordings.[27]

Compulsory licensing will be the eventual solution to the distribution of video programming online, and is already topping the policy debate in audio programming for streaming radio stations (Krebs, 2001). In Canada, greater progress has been made on compulsory licensing for video programming in the far larger broadband market per capita. Whereas Canadian webcaster ICraveTV was unable to prove its ability to prevent international reception of programming in 1999, and therefore lost its copyright arbitration and court case, JumpTV in 2001 is demonstrating far greater control over access to programming (Geist, 2000; JumpTV, 2001). As in other areas of video-over-IP, it appears that the more local the service, the greater the opportunity to work within existing regulation and property rights.

## 6. CONCLUSION: THE CHALLENGE FOR ANY-TO-ANY

For video-over-IP, the legal and regulatory issues arising in connection with government intervention is less about jurisdictional avoidance than global localization (see Reidenberg, 1999). "Information wants to be free," it was said in the early days of cyberspace. In the increasingly ubiquitous environment of the Internet in which commercial ISPs find themselves, information wants to be controlled by its owners and recipients. Building the Global Information Infrastructure is the largest engineering and capital development task ever faced. Video drives the future Internet because humans are visually literate far more than they can ever be intuitive consumers of text, graphics, or stand-alone audio. That future Internet is now being defined, by

---

[26]For a more general treatment, see Barton (1997).

[27]*Video on the Internet: Icravetv.com and Other Recent Developments in Webcasting: Hearing before the Subcommittee on Telecommunications, Trade, and Consumer Protection of the Committee on Commerce House of Representatives,* 106th Cong., 2d sess. 2 (2000) (testimony of Peter Jaszi, Professor of Law, Washington College of Law, American University). Retrieved from http://comnotes.house.gov/cchear/hearings106.nsf/a317d879d32c08c2852567d300539946/8d45454ad293f0db85256965006e67c1/$FILE/94.pdf

the standards bodies referred to in individual subsectors of the streaming video industry, but also by the Internet's future itself.

Those profoundly challenged by changing business models, notably broadcasters, copyright holders, and switched circuit telcos, will tell their governments and regulators to stop this market developing. In order for legacy property rights to be monetized, video-over-IP must initially be under national (even local loop) control. Critics are correct that this will curtail the end-to-end Internet until and unless it becomes standardized for profit-making rich-media applications. The development of property rights in broadband networks and services depends on such economic imperatives.

## ACKNOWLEDGMENTS

The 2000/1 research upon which this chapter is based would not have been possible without the support of colleagues at Re: Think! Consultancy, especially Ivan Croxford and Doug Laughlen. Especially generous interview and comment resources were given by staff and clients of the Independent Television Commission and British Broadcasting Corporation, streaming media and telecoms companies in northern Europe, and various academic and regulatory seminars in the period September 2000–July 2001. I wish to especially acknowledge the feedback from participants at the Harvard Information Infrastructure Project/Swiss Re conference at Ruschlikon- Switz. on June 28–30, 2001, and the Federation of Cinematic and Audiovisual Archives and Libraries (FOCAL) International Annual Conference in London on July 11, 2001. All errors and omissions remain my own responsibility. The information and sources in this chapter are accurate as of July 1, 2001.

## REFERENCES

Arbitron/Coleman Research. (2000, September). The broadband revolution. Retrieved from http://acw.activate.net/nab/092200/broadband.asx)

Bar, F., Cohen, S., Cowhey, P., De Long, B., Kleeman, M., & Zysman, J. (2000). Access and innovation policy for the third-generation Internet. *Telecoms Policy, 24,* 489–518.

Barton, J. H. (1997). The balance between intellectual property rights and competition: Paradigms in the information sector. *European Competition Law Review, 7,* 440–445.

Berkeley Law and Technology Center Conference. (2001). Berkeley, CA: Beyond Microsoft: Antitrust, Technology and Intellectual Property, Panel: Cable Open Access. Retrieved March 3 from http://www.law.berkeley.edu/institutes/bclt/events/antitrust/

Besen, S. M., & Saloner, G. (1989). *The economics of telecommunications standards setting.*

Blackman, C., & Nihoul, P. (Eds.) (1998). Convergence between telecommunications and other media: How should regulation adapt? [Special issue]. *Telecommunications Policy, 22*(3).

Burk, D. L. (1999). Virtual exit in the global information economy. *Chicago Kent Law Review, 73*(4), 943–995.

Carter Donahue, H. (2001). Opening the broadband cable market: A new Kingsbury commitment? Info, 3(2), Cambridge: Camford Publishing, (draft cited).

Cave, M., & Cowie, C. (1998). Not only conditional access: Towards a better regulatory approach to digital TV. *Communications and Strategies, 30* (2nd Quarter), 77–101.

Cherry, B., & Wildman, S. (2000). Preventing flawed communications policies by addressing constitutional principles. *L.REV. M.S.U.-D.C.L., 1,* 55–107.

Coase, R. H. (1937). The nature of the firm. *Economica, 4,* 386–405.

Commission of the European Communities. (1999). *Communications review: Recent developments in the market for electronic communications services within the EU and proposes certain possible future regulatory measures specific to this sector.* Retrieved November 10 from http://europa.eu.int/ISPO/infosoc/telecompolicy/en/comm-en.htm

Commission of the European Communities. (2001). *Microsoft agrees not to influence technology decisions of European digital cable operators.* Retrieved April 18, 2001. IP/01/569, from http://europa.eu.int/rapid/start/cgi/guesten.ksh

Compaine, B. (2001, January 12). *TV over Internet: Policies for convergence.* Paper presented at the Internet and Telecoms Convergence Consortium Members meeting, Massachusetts Institute of Technology, Cambridge, MA.. Retrieved from www.itel.mit.edu

Cowie, C., & Marsden, C. T. (1999). Convergence: Navigating through digital pay-tv bottlenecks. *Info, 1,* 53–66, Cambridge: Camford Publishing.

Croxford, I., & Marsden, C. T. (2001). *I want my WiFi! The opportunity for public access wireless local area networks in Europe.* London: Re: Think! Available at www.re-think.com

David, P., & Shurmer, M. (1996). Formal standards setting for global communications and information services: Towards an institutional regime transformation? *Telecommunications Policy, 20*(10), 789–816.

Eisner Gillett, S., & Lehr, W. (2000). The challenge of tracking broadband competition. Paper presented at the Internet and Telecoms Convergence Consortium, Massachusetts Institute of Technology, Cambridge, MA, at http://itel.mit.edu

Faulhaber, G. (2001). *Network effects and merger analysis: Instant messaging and the AOL-Time Warner Case.* London: London Business School. Retrieved from www.lbs.ac.uk

Federal Communications Commission. (2001). In the Matter of Amendment of Part 15 of the Commission's Rules Regarding Spread Spectrum Devices: Wi-LAN, Inc. Application for Certification of an Intentional Radiator Under Part 15 of The Commission's Rules, ET Docket No. 99-231 DA 00-2317, 10 May for release 11 May: http://www.fcc.gov/Bureaus/Engineering_Technology/Notices/ 2001/ fcc01158.pdf

Figueiredo, R. J. P., & Spiller, P. (2000). Strategy, structure and regulation: Telecommunications on the new economy. *L.REV. M.S.U.-D.C.L.*, *1*, 253–285.

Flynn, B. (Ed.) (2001). *Inside digital TV.* London: Philips Media.

Geist, M. (2000). iCraveTV and the new rules of Internet broadcasting. *University of Arkansas at Little Rock Law Review*, *23*, 123.

Gifford, D. J., & McGowan, D. (1999). A Microsoft dialog. *Antitrust Bulletin*, *44*, 619.

Gould, M. (2000). Locating Internet governance: Lessons from the standards process. In C. T. Marsden (Ed.), *Regulating the global information society* (pp. 193–210). New York: Routledge.

Grindley, P., Salant, D. J., & Waverman, L. (1999). Standards WARS: The use of standard setting as a means of facilitating cartels: Third generation wireless telecommunications standard setting. *International Journal of Communications Law and Policy*, *3*. Retrieved from http://www.ijclp.org/3_1999/ijclp_webdoc_2_3_1999.html

Grove, A. (2001, June 30). Keynote speech to Harvard Internet & Society conference.

Guilty: The appeals court's ruling shows that it was right to sue Microsoft. (2001, July 7). *The Economist.* Retrieved from http://www.economist.com/displayStory.cfm?StoryID=687513

Hugenholtz, P. B. (2000, April 2). *The great copyright robbery: Rights allocation in a digital environment.* Paper presented at New York University School of Law. Retrieved from http://www.ivir.nl/publications/hugenholtz/PBH-Ecology.doc

JumpTV. (2001). *Copyright board denies application to suspend Jump TV proceedings.* Retrieved June 4 from http://www.jumptv.com/mediaroom/pr2001-06-04.1.html

Kahin, B., & Abbate, J. (Eds.). (1995). *Standards policy for information infrastructure.* Cambridge MA: MIT Press.

Kahn, R. E., & Cerf, V. G. (1999, December). *What is the Internet (and what makes it work)?* Washington, DC: Internet Policy Institute, at http://www.internetpolicy.org/briefing/12_99.html citing October 24 1995 Resolution of the U.S. Federal Networking Council.

Krebs, B. (2001). Webcasters win right to join in royalties panel. *Newsbytes*, July 18, http://www.newsbytes.com/news/01/168095.html

Laffont, J. J., & Tirole, J. (2000). *Competition in telecommunications.* Cambridge MA: MIT Press.

Larouche, P. (1998). EC competition law and the convergence of the telecommunications and broadcasting sectors. *Telecommunications Policy*, *22*(3), 219–242.

Leiner, B. M., Cerf, V. G., Clark, D. D., Kahn, R. E., Kleinrock, L., Lynch, D. C., Postel, J., Roberts, L. G., & Wolff, S. (n.d.). *A brief history of the Internet.* Retrieved from www.isoc.org/internet/history/brief.html

Lemley, M. (2000). Will the Internet remake antitrust law? In C. T. Marsden (Ed.), *Regulating the Global Information Society* (pp. 235–242I). New York: Routledge.

Lemley, M. A., & Lessig, L. (2000, October 24). *The end of end-to-end: Preserving the architecture of the Internet in the broadband era.* Unpublished mimeo draft.

Lessig, L. (1999a). Brief as Amicus Curiae U.S. v. Microsoft, 65 F.Supp.2d 1 (D.D.C. 1999) (No. Civ. 98-1232 (TPJ), Civ. 98-1233 (TPJ) at cyber.law.harvard.edu/works/lessig/AB/abd9.doc.html

Lessig, L. (1999b). The limits in open code: Regulatory standards and the future of the net. *Berkeley Technology Law Journal, 14*(2), 759–770.

Levy, B., & Spiller, P. (1994). The institutional foundations of regulatory commitment: A comparative analysis of telecommunications regulation. *Journal of Law, Economics and Organization, 10*(2), 201–246.

Liebowitz, S. J., & Margolis, S. E. (1999). *Winners, losers & Microsoft: Competition and antitrust in high technology.* Oakland, CA: Independent Institute.

Marsden, C. T. (2000). (Ed.). *Regulating the Global Information Society.* New York: Routledge.

Marsden, C. T. (2001a). Cyberlaw and international political economy: Towards regulation of the global information society. *L.REV. M.S.U.-D.C.L., 1,* 253–285.

Marsden, C. T. (2001b). Property rights in the broadband space: A review of four conferences: Converging Communications 31 January 2001 (Global Communications Consortium, Regulation Initiative); Streaming Media Europe 12 October 2000 (Streamingmedia.com); TV Broadcasting Online 16 February 2001 (IBC Conferences); Ispconeurope.com February 2001 International Journal of Communications Law and Policy, Vol.6. Retrieved from http://www.digital-law.net/IJCLP/6_2001/ijclp_webdoc_14_6_2001.html

Marsden, C. T. (2001c). Shuffling the regulatory deck chairs. *Inside Digital TV.* Retrieved March 28 from http://www.re-think.com/ pdfs/ IDTV4-06.pdf

Marsden, C. T. and Ariñe, M. (2003). From analogue la digital. In A. Brown, and R. Picard (Eds.), *Digital Television in Europe.* Mahwah, NJ: Lawrence Erlbaum Associates.

Marsden, C. T., & Verhulst, S. (Eds.). (1999). *Convergence in European digital TV regulation.* London: Blackstone Press.

McGonagle, T. (2001, May). Does the existing regulatory framework for television apply to the new media? *IRIS Plus. Strasbourg: European Audiovisual Observatory.*

McGowan, D. (2000). The problems of the third way: A java case study. In C. T. Marsden (Ed.), *Regulating the global information society* (pp. 243–262). New York: Routledge.

Moe Terry, M. (1997). The positive theory of public bureaucracy. In D. C. Mueller (Ed.), *Perspectives on public choice: A handbook.* Cambridge: Cambridge University Press.

Naftel, M., & Spiwak, L. J. (2001). *The telecoms trade war*. Cambridge: Hart Publishing. Retrieved from http://www.phoenix-center.org/telindex.html

Noam, E. M. (1989). International telecommunications in transition. In R. W. Crandall & K. Flamm (Eds.), *Changing the rules: Technological change, international competition, and regulation in communications* (pp. 257–297). Washington DC: Brookings Institute.

North, D. (1990) *Institutions, institutional change and economic performance*. Cambridge: Cambridge University Press.

Reidenberg, J. (1999). Restoring Americans' privacy in electronic commerce. *Berkeley Technology Law Journal, 14*(2), 771–792.

Saltzer, J. W., Reed, D. P., & Clark, D. D. (1984). End-to-end arguments in system design. *ACM Transactions in Computer Systems, 2*(4), 277–288.

Samuelson, P., & Opsah, K. (1999). Licensing information in the global information market: Freedom of contract meets public policy. *European Intellectual Property Review, 21,* 386

Sandelson, D. (2001, February 16). Presentation at the IBC "TV Broadcasting Online" seminar, London. Retrieved from http://www.ibctelecoms.com

Shelanski, H.. (1999). The Speed gap: Broadband infrastructure and electronic commerce. *Berkeley Technology Law Journal, 14*(2), 721–744.

Tobias, M. (2000). *Streaming media Europe*. Keynote speech, October 11, London, Earl's Court. Retrieved from http://www.streamingmedia.com

U.S. v. Microsoft Corp. (2000). No. 98-1233 (U.S. District Court for the District of Columbia). *Findings of Fact,* pp. 39–41. Available at www.usd oj.gov/atr/cases/f3800/msjudgex.htm

Universal City Studios, Inc. v. Reimerdes (2000). 111 F. Supp. 2d 294 (S.D. N.Y. 2000).

Upgrading the Internet. (2001, March 22). *The Economist*. Retrieved from http://www.economist.com/displayStory.cfm?StoryID=S%26%28H %20%25Q%21%28%23%A

Waterman, D. (2001). The economics of internet TV: New niches vs. mass audiences. *Info, 3*(3), draft cited. (Cambridge: Camford Publishing).

Williamson, O. (1975). *Markets and hierarchies: Analysis and antitrust implications*. New York: The Free Press.

# 9

# Intellectual Property Concerns for Television Syndication Over the Internet

**Kenneth R. Carter**
*Columbia Institute for Tele-Information*

The Internet is generating a fundamental shift in how video content will be delivered and consumed. Previously, certain intellectual property rights were secured by means of a contract before a program was televised over any network (whether broadcast, cable, direct broadcast satellite, or the Internet). A key interest sought in these agreements was usually the exclusive right to televise a program to an audience in a specific geographic area. This system, based on granting exclusive intellectual property rights to a geographic region by means of a contract, will become obsolete as the new system evolves.

The structure of television distribution in the United States is based largely on the fact that electromagnetic waves that carry television signals only propagate through the ether for a limited distance before fading out so much that they cannot be received. The range of these radio waves is determined by a number of factors, including the curvature of the earth, atmospheric conditions, the height of the transmission tower, and the signal strength of the broadcaster. These physical limitations have shaped the development of the U.S. television industry. The ability to buy and sell exclusive program rights are an important feature of the system. However, the transition to Internet-delivered television will restructure this system by ignoring the segmentation of television program markets into geographic regions. Even if programs sold are cleared in blocks, such as station groups and owned and operated stations (O&Os), making that same program available for streaming or downloading over the Internet will fetter those rights tied to a particular geographic region.

Fortunately for the emerging system, the Internet is not hindered by the constraints of the old system. The Internet is not limited by geographic boundaries or by the imaginary lines used to establish and enforce markets. Because of its architecture, it is less distance sensitive to traffic than existing means of television delivery, such as broadcast or cable. For most Internet users, there is little difference between content from across the street and content from across the country. Moreover, the local networks that provide access to the Internet have less control over the access to content than do broadcast and cable TV networks. This chapter addresses the problems of adapting the existing system of video delivery, and the implications of this transition on related intellectual property rights. It will also examine how television distributors, by unwisely restricting content, have failed to take full advantage of the opportunities offered by the Internet.

"Reruns" of existing television programs will likely be a major component of the video content delivered over the Internet. This is largely due to the fact that the cost of airing a previously produced program is relatively small compared to the cost of producing new content for each airing. This provides a cheap source of content for an emerging medium that may lack the viewership to support new first-run content forms. Moreover, future syndication is an important anticipated revenue stream for currently produced programming because many shows are unprofitable until they achieve syndication.

Most rerun programs are distributed through television syndicators, so the rights to air these programs are currently tied up in numerous contracts that grant territorial exclusivity. A potential problem arises when new means of delivering content enables parallel distribution channels. These parallel channels disrupt existing intellectual property rights that grant exclusivity and whose underlying purpose is price discrimination. For instance, the syndicators of classic TV shows such as *M\*A\*S\*H\**, *Gilligan's Island,* or *The Jeffersons* may have restricted themselves from engaging in Internet distribution. They may have already promised a local station an exclusive program right and is therefore now unable to offer that program in the same region, albeit though a competing new technology. These intellectual property rights limit the availability of content to Internet TV.

## THE HISTORY OF TELEVISION

Two trends have characterized the development of the U.S. television industry. The first is an ever-increasing channel capacity at decreasing per-channel cost. This first trend enabled the second trend, which is the growth of the footprint of local networks that deliver an increasing quantity of content to a national audience. Initially, the footprint of local distribution was based on the distance limitation of signals (initially broadcast through the ether, then eventually over cables). Even today, 210 identifiable local U.S. television markets remain. This geographic segmentation will begin to dissolve as the local networks, which provide access to the Internet, loose their control over access to content.

### Early History of the Broadcast Networks

Television networks in the United States evolved from the radio networks established in the 1920s. The major attraction that drew local stations, in both radio and television, to affiliate with a network was the flow of network programming. Network programming drew in viewers to the station and consequently allowed the affiliate to gain from advertising sales revenues. Starting in the 1950s, these local stations began to align themselves with one of the three national networks: ABC, CBS, and NBC. The affiliates and the networks split the commercial airtime sold to commercial advertiser order to pay for the programming.[1] Local stations received broadcasts from other parts of the country, taking advantage of the scale of the combined content resources of the network.

Networks then took the step of applying the affiliation formula (once geared exclusively to radio broadcasts) directly to the television industry. Affiliates found the system even more beneficial with the television industry than they had with radio, primarily because the costs involved in the production of television programming were much higher.

The system has changed somewhat in that the networks often buy programming from outside suppliers instead of developing all their own programming or using entirely network programming. In the 1970s, the FCC's FinSyn rule prohibited the major broadcast networks from taking a direct financial interest in the syndication of its programs. This rule was repealed in 1991. Despite these changes in structure, the underlying principle remains the same: networks provide national television broadcasts to local stations in exchange for the use of airtime. This locked up the distribution of content in a limited number of powerful firms.

---

[1]CBS worked to entice affiliates to join its network with incentives regarding network compensation—the hourly fee paid by networks to the affiliates for airtime. Eastman, Susan Tyler. *Broadcast/Cable Programming*. Belmont: Wadsworth Publishing Company, 1993, p. 191. One enticement was to provide some network programming, for which the affiliates did not have to pay, in exchange for use of the station's airtime for sponsored programming. This system of cash and barter still prevails today.

## Cable Television

Cable television provided a step in and opened this local bottleneck. The inception of cable television began rather unceremoniously in the 1940s as community antennae (CATV)[2] systems. For many years, cable served exclusively as a means to improve reception between a city and its outlying rural areas. The need for reception enhancement was pronounced in various rural and mountainous areas, or where FM signals were blocked by natural obstructions such as hills and the curvature of the Earth. The use of cabling to carry signals diminished the effect of obstructions on the connection between a broadcast signal's origin and the intended recipient. Initially, with one antenna, a cable system could serve a building, or even a small neighborhood. Eventually this system expanded to provide commercial service over a larger area.

The foundation of today's cable system emerged out of this framework. In the 1950s, the system took another step forward. Eventually, cable systems began to use microwaves as a means of transmitting broadcasts from cities to rural areas. The goal of this effort was to increase the distances that the signals could travel. Soon cities in rural areas could import broadcast transmissions from distant urban centers. This had the effect of greatly increasing the number of programming choices for the rural inhabitants. The ability to receive broadcasts from these distant stations meant that viewers could choose from a larger variety of programming, one that was less dependent on geographic location.

Pay TV added a whole new dimension to the cable television industry. Home Box Office (HBO) emerged in 1972 and was the first successful pay TV service. In order to broadcast both locally and nationally, HBO first employed broadcast over microwaves and then began to use geosynchronous earth-orbiting satellites. Over the following decades, a slew of new channels designated for specific topics, such as sports and movies, were created. This allowed multiple system operators (MSOs) to offer tiered packages of programming from which the viewer could pick (Vogel, 2001). This marked the first time that programming was not at all linked to local affiliates' broadcasts. However, the distribution of content remained tied to local MSO conduits.

## Satellite Television

The use of satellites to deliver television signals directly to viewers made it possible to expand beyond the limited distances covered by broadcasting

---

[2]Today, it is a misnomer that CATV stands for the CA in cable television. The CA is more appropriate for community antennae.

and cable. During the 1980s, individual viewers could, for the first time, cheaply obtain satellite dishes to receive television. This technology provides two different services: broadcast and interactive. Broadcast services include digital video broadcasting (DVB), normal television choices, near video-on-demand, pay-per-view, and data broadcasting. Interactive services such as televoting, online shopping, gaming, and so on[3] use a two-way connection between the user and the service provider. DBS from DirecPC, for instance, can also offer Internet services allowing connection speeds around 400 Kbps, provided that the receiving dish is in view of the southern horizon. This direct connection between the end user and the service provider, without the need for a cable system or any other geographically based medium, is part of the trend to overcome geography as an obstacle to the reception of broadcast signals.

## Television Distribution Today

As a result of digital cable and satellite television, both programming and delivery are no longer tied to geography. Despite this technological shift, however, the regional nature of television remains alive and well. According to the Federal Communications Commission, there were 1,288 VHF and UHF television stations in the United States in September 2001. The signals from each of these stations can only be received within an area of 60 miles from the transmitter, and maintaining local television markets. According to Nielson, there are presently 210 television markets in the United States in which local television stations, local cable systems, advertisers, and syndicators buy and sell commercial programming.

The Mass Media Bureau of the FCC is charged with the regulation of over-the-air broadcast television stations. The FCC performs this task by assigning frequencies, operating power, and granting exclusive territory licenses to stations. Assignments are determined by a number of factors, including the curvature of the earth, atmospheric conditions, and the signal strength of the broadcaster. The FCC enforces these rules by issuing broadcast licenses to stations for a period of only 8 years. Therefore, if a licensee fails to comply with statutes or FCC rules and policies, then the FCC may refuse to renew its license. However, because this would constitute a death sentence for the station, these punitive measures are rarely imposed. Accordingly, the issuing of new licenses is extremely rare. At the time of writing, the Mass Media Bureau was not accepting applications for

---

[3]SMATV Systems Enhanced with Satellite Based Interaction Channel.

new broadcast licenses. This limits the amount of available over the air channel capacity.

## Television Distribution

Copyright law provides the basic framework for how television content is distributed. The copyright is a bundle of distinct rights that is the primary legal construct for protecting creative expression. A copyright gives the owner certain exclusive rights in an artistic audio or visual work, and is designed to stimulate the production of such works by enabling creators to receive compensation and credit for the use of their work. Once the copyright expires, these rights fall into the public domain, and are available for general consumption free of legal restraints on duplication and distribution. In order to obtain copyright protection, one must create an original work of authorship that is fixed in a tangible medium of expression.[4] Section 106 of the Copyright Act of 1976[5] established five exclusive rights that are related to video program distribution. Most notable of these for IPTV are the rights of reproduction, distribution, and performance.

Distributors do not sell an actual program to local stations in the existing 210 markets; rather, these stations are sold their broadcast rights to the program. A station usually purchases exclusive rights to broadcast the program in its market area so that it can maximize viewership and the station's earnings. In cases where the network buying the programs has national coverage, direct sales to these markets are usually preempted.

Syndication is a crucial factor in shaping the production process, even at its earliest stages. Because networks are much more willing to produce shows that have good syndication and rerun possibilities, shows that demonstrate this potential are more likely to be approved. As television shows almost inevitably go into debt upon the start of their production, the terms reached through contract negotiations between the network and the producers of a show are crucial to determining how a show will be produced. Many times, in exchange for funding the initial costs of production, networks will seek to obtain from producers a network options clause. These clauses give the networks the right to order shows for a given amount of time, thereby fixing a minimum number of shows and securing a minimum commitment from the production staff. Another such network option that may be used reserves the right of the network to reject any episodes produced after this minimum number of shows has been reached. Additionally, these two types of clauses are often used together. As a result, if another network were to give the producers a better offer, such an offer would violate these contract terms and could not be ac-

---

[4]17 U.S.C. § 206.
[5]17 U.S.C. §§ 100 *et. seq.*

cepted. These conditions also benefit the original network by allowing it to keep the show on its lineup for a price that is below the fair market value of the show. The network justifies retaining this right by asserting that it took the initial risk in funding the original production of the show (Vogel, 2001). Therefore, the network is rewarded for taking this high, early risk by being in a position to keep a successful show on its airwaves, and away from its competitors, at a relatively low price.

A contract with one of the networks can increase a producer's chance for securing funding from banks or other institutions. When a contract with the network expires, after the initial run or after reruns, pursuant to the contract, the show's ownership vests in the producers, and they are free to market the program as they like. Shows that last three full seasons with good ratings have the most potential for syndication. Once producers own the rights to the show, they can sell numbers of episodes to local television and cable stations for resyndication broadcast. These shows will often run daily over an extended period, although this depends on the number of episodes for which the station paid. Syndication market licenses are generally sold to the highest bidder and tend to range between 3 and 6 years, but recent trends have leaned toward the shorter time frame.

First-run syndication is another option for producers. Rather than setting up a normal network contract that has long-term ramifications, the producers design a show to be sold to nonnetwork affiliates for their first appearances. This short-term arrangement is an opportunity for local stations and other affiliates to fill the airtime that precedes the network evening programming. These shows are low cost, and include game shows, talk shows, and tabloid news shows. First-run syndicated shows are nonnetwork options that are relatively inexpensive and suit the needs of local stations to fill the nonnetwork airtimes. These shows do not depend on popularity or a long running life; rather they are inexpensive short-term shows that entirely skip the networks themselves (Vogel, 2001).

### Piracy and Syndication

One primary concern of television distributed over the Internet is that of illicit copying. Digital media affords the opportunities to make and distribute perfect copies of video programs. To misquote Shakespeare, these new technologies evoke the desire, but inhibit the performance.[6] It is unlikely that video pirates will run rampant on the Internet. This is true for a variety of reasons, beginning with increased legal protection. Duplication costs are not only inexpensive for pirates, they are also inexpensive for distributors who can lower their prices or offer more value added services. In addition, the technology available to protect against unauthorized duplica-

---

[6]Shakespeare, William, *MacBeth*, Act II, Scene III, "... it provokes the desire, but it takes away the performance ..."

tion is growing more pervasive. As Einhorn (chap. 10) suggests, it is unlikely that there will be room for compulsory licensing. Moreover, the previously widespread defense of fair use is likely to be curtailed. In fact, a shift in the current laws regarding encryption has acknowledged the lack of fair use defense, and has even gone so far as to preempt it.

One scenario for the rampant copying of television content over the Internet may be a type of video Napster. Currently, this infamous site has been almost entirely shut down;[7] however, the concept remains as new sites parallel in concept are cropping up on the web,[8] where software enables peer-to-peer sharing by users of MP3 music libraries. Successful technologies for recording and reproducing music are often leading indicators for those technologies that might be adopted for the more memory intensive video content. The not too distant future of television distribution may be witnessed by sites such as Morphius, a peer-to-peer community of users trading video files, TV programs, and movies over the Internet using a Napster-esque arrangement.[9]

A new feature of TiVo allows subscribed users to "e-mail" TV shows to other subscribed users. This may provide the content distributor with an increased audience, an important measure for advertising supported programming. The TiVo is not an open system like a personal computer (PC), so it will be easier for networks to keep track of its audience and even to charge the user directly for content. Who will extract the most benefit from this is simply a question of bargaining power. It is likely to go to the major syndicators.

Distributors' success depends on their ability to control postsale copying; they may construct a number of technological and legal obstacles to counter would-be infringers. A key strategy for distributors of video content over the Internet is to use a diverse mix of technical, business, and legal measures that change from product to product and from release to release. This series of safeguards is an effective deterrent by forcing would-be infringers to run a gauntlet of obstacles to pirate the work. The varying of protective measures has another advantage. The knowledge acquired by a pirate in a previous successful defeat of a specific safeguard

---

[7]The RIAA sued Napster and obtained an injunction effectively shutting down Napster for promoting music piracy through the encouraging illegal copying of copyrighted music. In 2000, the German media giant Bertelsmann concluded it was unwise to fight the underlying file sharing technology of Napster and the clear demand it has created on the Internet. It received an option to buy a stake in the company in exchange for a loan to help Napster change its service from a free file sharing software into a subscriber-based business and it will begin paying record companies.

[8]For a means people are using to circumvent the district court's order, see http://www.napcameback.com . The site uses encryption to allow Napster users to circumvent copyright filters on Napster.com

[9]http://www.morphius.com

will not automatically pay dividends in bypassing a different safeguard. Some of the technology responses to piracy are security and integration of operating systems, file access, rights management language, encryption, watermarking, access control, marking and monitoring, sniffer technologies, copying function alerts, noncopying embedded passwords, source identification (SID) codes, bar codes, and virus seeding.

Another strategy to combat infringement employed by distributors of video content over the Internet is to change their business strategy in response to piracy. A seemingly obvious response is to lower prices (National Research Council Computer Science and Telecommunications Board, 2000). Additionally, distributors could modify their products to make copying by pirates more costly (Shapiro & Varian, 1999). If the cost of reproduction or piracy is high relative to the cost of acquiring work legitimately, then pirates are likely to be deterred from infringing. This is the case with media like newspapers, magazines, and paperbacks. Alternatively, the product distribution strategy may also be modified to safeguard against infringement. For example, a distributor might elect to switch from downloading to streaming technologies in order to frustrate pirates.

A strategy more attuned to computer hardware and software than to entertainment content is to speed up new versions when a "clone" enters the market. An online firm can link with a physical product by offering online content, thereby increasing the sales of physical versions (Fisher, 2000). By adding value to online information, these online versions are more desirable than hard copies. That is to say, video distributed over the Internet should not be just video online. Distributors should seek to add elements that surpass VHS or DVD versions.

The defining characteristic of an information good is that it qualifies as an "experience good"—that is, consumers do not know what it is worth until they experience it (Shapiro & Varian, 1999). Reduced distribution costs enable increased advertising through the distribution of free samples, because it is easy to give away something that has zero marginal cost of distribution. The strategy behind this advertising scheme is to divide a product into components that are given away and that are sold. Give away only part of a product as a free sample to sell similar, but not identical, products. The theory is that by providing free samples of a product to the marketplace, demand for that product will be stimulated. For example, the full text of books and reports available online often increase sales of hard copy versions of those works. The Internet also provides free access to small pieces of large products like encyclopedias and databases but that are too difficult to reassemble into their comprehensive form. However, this access entices consumers and in many cases leads them to purchase the hard copy versions they have used online.

Over the past several years, federal protection of intellectual property rights has grown, and criminal sanction for infringement has been reinforced with new criminal provisions. New legislation has included the following: Copyright Infringement Act, the Computer Fraud and Abuse Act, and the Economic Espionage Act (1996).[10]

In addition, nearly every U.S. state has enacted some form of computer-specific legislation. Most notable is the No Electronic Theft Act of 1997[11] ("NET Act"). The NET Act has fine-tuned some common law definitions concerning infringement. Traditionally, it was difficult to prosecute small, personal use infringers. The NET Act, however, provides effective recourse against small-scale, willful, copyright violators who are not motivated by "commercial interests." The act authorized criminal prosecution for making merely 10 illicit copies of a protected work, worth just $2,500. These protections have sought to limit postsale illegal copying, but also carry with them the unintended consequence of severely limiting fair use of copyrighted materials. Some have also argued for a mandatory licensing scheme for distribution by means of a Napster framework, but this is unlikely (see Einhorn, chap. 10 in this vol.).

It is impossible to sue every infringer, and even if it were plausible, it is not a smart business move for a company to make a habit of suing its own customers. Technical solutions are not likely to exhaustively counter infringers, because they are not likely to be cost effective on a large scale.[12] The best solution to piracy protection is a diversity of protection measures and distribution channels. This, of course, should be tempered by the fact that features should only interfere minimally with user's enjoyment of the product.

## INTERNET AND BROADCAST TV

### Drivers of New TV Technologies

Internet television is not an invention that is likely to catch on immediately. Rather, it will likely grow in popularity as new technologies, content formants, and business models are gradually adopted. As the other contributors to this volume suggest, there already exist several new formats that run a continuum for enhancement, such as interactive program guides to downloadable or streaming video. The adoption of new television technologies such as cable, the VCR, and satellite distribution have been driven by two key factors: reruns and pornography. (The latter is not discussed here.) Reruns of previously aired programs serve the needs of both

---

[10]*See* 18 USC 90.

[11]*See* HR. 2265, P.L. 105-147, 111 Stat. 2678—codified in Title 17 and Title 18 of USC.

[12]Were cost not a factor, encryption technology might be a viable solution. For one thing, encryption scrambles content so that it cannot be unscrambled or transferred to another device without the correct software key. Also, CD burners add a digital serial number to every CD they copy, which enables each copy to be traced back to the individual machine.

supply and demand, and content is needed to fill the ever-increasing channel capacity of video delivery systems.

In 1992, Bruce Springsteen sang, "Fifty seven channels and nothing on." A decade later he would only be a quarter right. Today, there are over 200 channels offered by broadcast, cable, and satellite (Noam, 2001). This falls far short of the much-promised 500-channel universe. In New York City, not known as a leading market in cable channel capacity, total program capacity is over half a million program hours per year, having grown at a compound annual rate of over 10% for the last 30 years (Noam, 1998). However, this decreasing cost of capacity has created an incredible demand to fill the shelf space of television distribution networks.

The new shelf space has provided the opportunity for more programming content. Over the last decade, new channels have sprung up on a seemingly weekly basis. The channels are increasingly more specialized and offer a broader array of topics. This has given rise to the concept of *narrowcasting*, a term that describes the idea of broadcasting to a narrow audience. Narrowcasting generally targets audience shares of less than 1%. Through the loss of scale economies, narrowcasters must make up margins by offering their audiences specialized and therefore more valuable programming content. Taking this just one step further, many authors have envisioned customized and individualized programming over the Internet to still narrower groups of viewers (or to one viewer) in the near future (Noam, 1994). Consequently, Internet Delivered Television, or Internet TV, has evolved.

Hart (chap. 14, in this vol.) identifies six categories of content models for Internet TV. He suggests that the distinction between what is considered new or old content is hard to determine. This chapter primarily concerns itself with only three aspects of distribution that have been made possible by Internet delivery options: the programming of local TV stations, syndicators, and licensors of web video.

Local television stations are already employing streaming video on the Internet to extend their reach to audiences, primarily for local news programming of local network television affiliates. In the future, however, local stations may also decide to make other forms of content available. Further upstream, syndicators are now able to offer programming directly to consumers. A syndicator can now move beyond brokering deals at the National Association of Television Production Executives (NATPE) to offering streaming or downloadable content directly to end-users. So, instead of having to wait for a rerun of an episode of *Gilligan's Island,* the episode could be available immediately, at the consumer's request.

### Inhibitors of New TV Technologies

This fundamental change in the way that video content will be delivered and consumed will redefine the current notions of intellectual property rights and territorialism. The old system based on exclusive intellectual property rights granted by contract will be overtaken by the new system. This will largely result from the fact that the Internet does not follow the same rules as broadcast or cable, because Internet transmission is not sensitive to distance.

However, this transition cannot occur overnight. The immediate problem is that much of the existing content is tied up in distribution contracts. These contracts grant the exclusive rights to the property for specified periods of time and to limited geographic areas. This prevents syndicators from distributing their product over a different and competing media such as over the Internet. Such centralization denies the existing right of different broadcasters to provide available content across geographic boundaries. So how is a syndicator to take advantage of the cutting edge means of video delivery? It is unlikely that the participants will abandon the current system in a wholesale fashion, so a single strategy will not be viable.

## Lucas in Love

The unavailability of traditional content for delivery over the Internet may ironically drive new formats. Without reruns to fill the channel space, web-based distributors will have to finance new programs. In light of the recent wave of dot-com failures and the tightening of capital budgets, this may be difficult in the near future. Addressing the crucial issues of distribution and fair use, a particularly interesting intellectual property case is the 1999 short film *George Lucas in Love*.

*George Lucas in Love* is a short film intended as a calling card, or résumé, by USC Film School graduate student Joe Nussbaum. This "web-short" is a parody of both the 1977 George Lucas classic *Star Wars* and the 1998 *Shakespeare in Love,* starring Gwyneth Paltrow. The film commences as George Lucas, a young USC Film School student, is unable to complete his thesis due to writer's block. Lucas desperately needs to finish his screenplay in three days in order to graduate from school. And, à la the protagonist in *Shakespeare in Love,* Lucas does not notice the potential material in his surroundings that could act as a basis for his film. His film school world is filled with inspiration for what will one day be the eminently recognizable characters from the *Star Wars* trilogy. Yet he fails to view his experiences as material for his film. In despair, and at the brink of destruction, Lucas happens to meet a lovely young co-ed, a doppelganger for Princess Lea, who helps to inspire him to see the potential around him and to complete his film. Whereas *George Lucas in Love* was masterful, unfortunately for writers Joe Nussbaum and Daniel Shere and for producer Joseph Levy, the National Academy of Motion Picture Arts and Sciences ruled that web-shorts are ineligible to be nominated for Oscar awards.

Despite being Oscarless, *George Lucas in Love* has become one of the most downloaded short films on the Internet. Interestingly enough, it is available through two seemingly competing Internet distribution channels. It is available for download via streaming media from mediatrip.com for free. Alternatively, it is available on DVD and VHS at Amazon.com for $12.99 and $7.99, respectively. Trying to sell something that is also being given away for free does not seem like a viable business model. However, it appears that anything is possible on the Internet. In fact, *George Lucas in Love* has been one of Amazon.com's top sellers; it even outsold the *Phantom Menace* in its first month.

This case study demonstrates how the physical and ephemeral can coexist. Just as the radio broadcast of a song is an ad for itself, as well as a substitute, this means of video distribution turns out to be a free sample (Shapiro & Varian, 1999). Because the presentation over the web does not come in an easily accessed, consumer-friendly format, the two distribution channels can coexist, thereby increasing and not cannibalizing demand.

## CONCLUSIONS

The central problem is one of mind set. The syndicator has to see the new opportunities that Internet TV affords. However, what is the syndicator to do about preexisting content contracts and relationships? Nothing? What can be done is to take advantage of the new technologies. One such opportunity is to limit remote access to content. This can be done by technologies blocking delivery to server IP addresses for specific geographic locations, whenever possible. Another approach is to use server caching technologies such as Akamai. This limits content availability only in geographic regions already tied up with syndication contracts. That might solve the problem going back. Going forward, the syndicator may want to consider not tying up content with geographic exclusivity contracts. Nonexclusivity may, in fact, generate more revenue for all distribution channels. Through a diversity of protection measures and parallel, noncompeting distribution channels, program distributors can continue to price discriminate among the end-users of its products. This will ensure maximum revenue for program sources amid changing distribution and business models.

## ACKNOWLEDGMENT

I would like to acknowledge the help of Uriel Cohen and Brian Bebchick in preparing this chapter.

## REFERENCES

Botein, M. (1980). *Network Television and the Public Interest*. New York: Lexington Books.

Eastman, S. T. (1993). *Broadcast/Cable Programming*. Belmont: Wadsworth Publishing Company.

Fisher, W. (2000, October 10). *Digital Music: Problems and Possibilities*. Prepared for A Free Information Ecology in the Digital Environment, New York University Law School, March 31, 2000.

National Research Council Computer Science and Telecommunications Board. (2000). *The Digital Dilemma: Intellectual Property Rights in the Information Age*. Washington, DC: National Academy Press.

Noam, E. (2001, June 27). "Two cheers for the commodification of information." *Journal of Intellectual Property*, p. 5.

Noam, E. (1998). "Public Interest Programming in American Television." In E. Noam & J. Waltermann (Eds.), *Public Television in America* (pp. 145–175). Gutersum, Germany: Bertelsmann.

Noam, E. (1994). "The Stages of Television: From Multi-Channel Television to the Me-Channel." In C. Contamine & M. van Dusseldrop (Eds.), European Institute for the Media (pp. 49–58).

Owen, B. M. (1999). *The Internet Challenge to Television*. Cambridge: Harvard University Press.

Shapiro, H., & Varian, C. (1999). *Information Rules*. Cambridge, MA: Harvard Business School Press.

Vogel, H. L. (2001). *Entertainment History Economics*. New York: Cambridge University Press.

# Internet Television
# and Copyright Licensing:
# Balancing Cents
# and Sensibility

Michael A. Einhorn*

*Voice mail: 973-618-1212. email: meinhornphd@hotmail.com The author wishes to thank Jane Ginsburg, Assaf Litai, and Robert Pepper for helpful comments. This paper appeared in the 20 CARDOZO ARTS AND ENTERTAINMENT LAW JOURNAL 2 (2002) and is reproduced with the express permission of the journal.

## INTRODUCTION

In a speech delivered in October 2001 to the National Summit on Broadband Deployment in Washington, D.C., the Federal Communications Commission ("FCC") Chairman Michael K. Powell stated:

> Much of what is holding broadband content back is caused by copyright holders trying to protect their goods in a digitized environment (in other words, a perfect reproduction world). Stimulating content creation might involve a re-examination of the copyright laws. Arguably, VCRs would not be widely available today if Universal Studios had won its infringement case against Sony in 1984.[1]

---

[1]Michael K. Powell, Chairman, Federal Communications Commission, Remarks at the National Summit on Broadband Deployment, Washington, D.C. (Oct. 25, 2001).

Though the Chairman's remarks made no specific recommendations, a possible area for further consideration would be retransmission rights for local television signals that can be captured and re-sent over the Internet.

Internet television would entail a new distribution technology that could enable video content to be transmitted to personal computers or digital set top boxes that interface with the Internet protocols (a.k.a. TCP/IP). It would present greater opportunity for viewer interactivity, user editing, and the personalization of advertising. Internet distribution should not be expected at the outset to transform content greatly, although some niche programming and off-network distribution can reasonably be expected. As had been the case with terrestrial cable in the 1970s, emerging video applications that enhance the distribution of content may "jump start" the base of broadband users, and provide economic support for further investments in high-quality content. This could lead to more complete transformations of content and integration of technology and video product.

Digital and Internet technology can enable the following new capabilities:

*Time-shifting:* Users may view programs at more convenient times.[2]

*Space-shifting:* Users may view appealing content in more convenient locations, such as those enabled by wireless technology.[3]

*Personalization:* Providers may insert personalized ads and provide video material to users that are more tailored to individual tastes, as revealed by online behavior.[4]

*Screening:* Video providers may strip programs of content unsuitable for children, per the personalized instructions of the receiving home.

*Transforming:* Providers may "cut and paste" segments from different shows for edited viewing.

*Multimedia:* Providers may combine different works (e.g., video and music) for simultaneous presentation.[5]

---

[2]*See* Roxio Software, *at* http://www.mgisoft.com/products/mgitv/ (last visited Jan. 24, 2002).

[3]*See* WC3 Synchronized Multimedia, *at* http://www.w3.org/AudioVideo/#Background (last visited Jan. 24, 2002) (explaining multimedia combination).

[4]*See* Net Perceptions, *at* http://www.personalization.com (last visited Jan. 24, 2002) (offering a website with commercial services).

[5]*See* WC3 Synchronized Multimedia, *at* http://www.w3.org/AudioVideo/#Background (last visited Jan. 24, 2002) (explaining multimedia combination).

*Morphing:* Characters and designs may be digitally transformed in creative manners that add new dimensions or ideas to the material.[6]

*Archiving:* Content may be archived on servers for subsequent viewing.

*Repackaging:* Content can be represented in different venues; e.g., a web site can combine programs from different sources that have a common theme.

*Hyperlinking:* Viewers can surf and skip from video content to related links about particular items in the program.[7]

*User Communities and Chat Rooms:* Users may establish cyberclubs regarding particular content items that most interest them.[8]

Not all broadcast television signals can present fair game for free takings by Internet retransmitters. Evidently, capture and retransmission present a potential danger to copyright owners in broadcast programming. For digital technology, secondary users may make and distribute near-perfect copies of broadcast material. Without proper copyright authorization, Internet technology could then distress program investments, and reduce financial incentives to provide or distribute new content.[9]

To expedite the copyright process, several Internet service providers (including America Online before its acquisition of Time Warner) unsuccessfully lobbied Congress in November, 1999 to grant rights for reuse of television signals, to be compensated via compulsory licensing.[10] If compulsory licensing were enacted, cyber-providers would be able to use, without direct owner authorization, copyrighted program material with

---

[6]*See* MIT Artificial Intelligence Laboratory, *at* http://www.ai.mit.edu/people/spraxlo/R/superModels.html (last visited Jan. 24, 2002) (illustrating morphing).

[7]*See* LinkBaton, *at* http://my.linkbaton.com (last visited Jan. 24, 2002).

[8]*See* InfoTreks, *Best Chat Room List, at* http://www.infotreks.com/chat.html (last visited Jan. 24, 2002).

[9]These dangers became headline news in February 2000, as a coalition of American television broadcasters successfully enjoined and negotiated the cessation of unauthorized retransmissions by iCraveTV, a Toronto-based Internet company that picked up and retransmitted signals from seventeen American television stations. *See* Dugie Standeford & John T. Aquino, *Internet Broadcasting; U.S. Studios Win Injunction Against iCraveTV,* Internet Newsletter, Feb. 2000, at 3.

[10]*See* Patricia Fusco, *AOL Lobbies for License to Carry Local TV Stations, at* http://www.internetnews.com/isp-news/article/0,,8_236121,00.html (last visited Jan. 28, 2002) (stating that statutory permission was to be introduced in the Satellite Home Viewer Improvement Act of 1999, Pub. L. No. 106-113, 113 Stat. 1501, 1536).

statutory fees determined under the jurisdiction of the U.S. Copyright Office.[11] Internet video providers could then provide access to popular content without having to track down and negotiate deals with copyright owners. Congress held subsequent hearings in June 2000 on the matter.[12]

However, the hearing's subcommittee found that the information requirements for compulsory licensing of Internet retransmissions were inappropriate for the wide diversity of uses and geographic dispersal of the potential viewing community.[13] If compulsory licenses were designed to compensate for potential economic loss, it would be necessary to determine how many original viewers would be lost to a particular retransmission of a program to an Internet audience.[14] Displacement ratios can vary considerably among different applications and geographic regions. Furthermore, any administrative or statutory formula, once established, is likely to be inflexible as economic conditions change.[15]

Rather than mandate compulsory licenses, an alternative strategy would exempt certain limited uses of television programs broadcast over free radio spectrum. This could be made possible through voluntary agreement or, more arguably, by statute.[16] Following imperfectly the three-part fair use paradigm set out by Wendy Gordon,[17] exemptions may

---

[11]*See* 17 U.S.C. § 801 (2000).

[12]*See Copyrighted Webcast Programming on the Internet: Hearing Before Subcomm. on Courts and Intellectual Prop. of the House Comm. on the Judiciary*, 106th Cong. (2000), *available at* http://www.house.gov/judiciary/courts.html (last visited July 8, 2000).

[13]*See id.* at 30. The Committee explained: Our principal concern is the extent to which Internet transmissions of broadcast signals can be controlled geographically. The Internet is a worldwide system with the capability of transmitting, or retransmitting, copyrighted works to hundreds of millions of viewers within seconds. If a compulsory license were created for retransmission of local broadcast signals, it is unclear how the retransmission of those signals could be limited to their local markets. *Id.*

[14]*See Statement of the Register of Copyrights: Hearing Before the Subcomm. on Courts and Intellectual Prop. of the House Comm. on the Judiciary*, 106th Cong. 47 (2000) (statement of Mary Beth Peters), *available at* http://www.house.gov/judiciary/courts.html (last visited July 8, 2000); *see also* U.S. Copyright Office, A Review of Copyright Licensing: Retransmission of Broadcast Signals 92-100 (1997), *available at* http://www.loc.gov/copyright/reports (last visited Jan. 24, 2002).

[15]*See* Stanley M. Besen et al., *Copyright Liability for Cable Television: Compulsory Licensing and the Coase Theorem*, 21 J.L. & Econ. 67, 68 (1978).

[16]We here take Wendy Gordon's point: "From the point of view of copyright owners ..., a system that permitted certain limited uncompensated takings to occur, as long as they did not cause substantial injury, might be preferable to a system in which compensation was guaranteed but only after the fact." Wendy J. Gordon, *Fair Use as Market Failure: A Structural and Economic Analysis of the Betamax Case and its Predecessors*, 82 Colum. L. Rev. 1600, 1623 (1982).

[17]*See id.* at 1614.

be reasonable when the transactions cost of licensing are high, an important public interest is served, and/or when the sale of advertising or programming is promoted.[18]

## COPYRIGHT, FAIR USE, AND ECONOMIC HARM

Copyright is federally protected by the Copyright Act of 1976 ("Copyright Act"), which became fully effective on January 1, 1978.[19] Section 106 established five rights that relate to the protection of video entertainment: (1) the right to reproduce the work; (2) the right to prepare derivative works based on the original; (3) the right to distribute copies of the work; (4) the right to perform the work publicly; and (5) the right to publicly display the work.[20]

Section 107 of the Copyright Act[21] codified the preexisting judicial doctrine of "fair use," which is a "privilege in other than the owner of a copyright to use the copyrighted material in a reasonable manner without his consent...."[22] Statutory factors to be considered in determining whether the use of a work is "fair" include: (1) the purpose and character of the use (duplicative vs. transformative; commercial vs. non-profit); (2) the nature of the original work (rote vs. creative); (3) the amount and substantiality of the use (partial vs. complete copying); and (4) the effect of the use upon the potential market or value of the work.[23]

More often than not, courts are reluctant to uphold a "fair use" defense when original content is creative, copyright holders are directly harmed, and copying is duplicative, commercial and/or complete. Included in the measure of market harm are foregone direct sales and lost opportunities to license content to users in existing or potential markets.[24] These considerations should affect any balanced discussion on copyright exemptions for retransmitted programs.

---

[18]See id. at 1601, 1618-21.

[19]17 U.S.C. § 101 et seq. (2000).

[20]See id. § 106(1)–(5).

[21]See id. § 107.

[22]Rosemont Enter., Inc. v. Random House, Inc., 366 F.2d 303, 306 (2d. Cir. 1966), cert. denied, 385 U.S. 1009 (1967) (quoting Horace Ball, The Law of Copyright and Literary Property 260 (1944)).

[23]See 17 U.S.C. § 107; see also Melville B. Nimmer, Cases and Materials on Copyright and Other Aspects of Entertainment Litigation § 13.05 (4th ed. 1991).

[24]See Harper & Row, Publ., Inc. v. Nation Enter., 471 U.S. 539, 568-69 (1985); Twin Peaks Prod., Inc. v. Publ'ns Int'l, Ltd., 996 F.2d 1366, 1377 (2d Cir. 1993); United Tel. Co. of Missouri v. Johnson Publ'g. Co., 855 F.2d 604, 610 (8th Cir. 1988); DC Comics, Inc. v. Reel Fantasy, Inc., 696 F.2d 24, 28 (2d Cir. 1982).

Once regarded to be most important, the fourth criterion provides an immediate opportunity for a segue into economic reasoning.[25] From an economic perspective, a reproduction or transmission of a work, now or in the future, may possibly displace or promote the direct sale of an original work, or interfere with the right of the owner to license its material. The economic importance of displacement and promotion is generally recognized in U.S. copyright law. For example, § 114 of the Copyright Act recognizes that certain digital audio transmissions of sound recordings may promote record sales, and therefore exempts from copyright protection performances on digital broadcast radio.[26] In a similar fashion, § 110 exempts performances of musical compositions that occur within the physical confines of record stores.[27] In negotiations regarding licensing fees for reproductions of musical compositions in digital media, the contending parties recognized that digital downloads may displace original CD sales, and adopted identical fees for licensing secondary reproductions in each.[28]

## SIGNAL RETENTION

As a result of two Supreme Court decisions, unedited over-the-air television signals in the U.S. may now be captured and transmitted for reuse by local and distant cable operators, with no need to compensate original station broadcasters.[29] The Court determined that cable operators are not so much broadcasters that engage in public performances of copyrighted programs, as they are passive recipients of material broadcast by others.[30] The basic function of their equipment is little different from that owned by a television viewer.[31] Accordingly, cable operators, "like viewers and unlike broadcasters, do not perform the programs that they re-

---

[25]See Harper & Row, 471 U.S. at 569. The U.S. Supreme Court had characterized the market harm as "undoubtedly the single most important element of fair use." Id at 566. However, one subsequent Court decision explored the four together and not in isolation. See Campbell v. Acuff Rose Music, Inc., 510 U.S. 569, 576 (1994). This modification was made to consider the transformative nature of parody to a copyrighted song. See id. at 570.

[26]See 17 U.S.C. § 114(1); see also Agee v. Paramount Comm., Inc., 59 F.3d 317, 320 (2d Cir. 1995).

[27]See 17 U.S.C. § 110(7).

[28]The compulsory license is established for secondary uses only. See id. § 115(1). Songwriters and music publishers retain exclusive copyright for the first recording of a copyrighted work. See Mechanical and Digital Phonorecord Delivery Rate Adjustment Proceeding, 64 Fed. Reg. 6221, 6226 (1999).

[29]See Teleprompter Corp. v. Columbia Broad. Sys., Inc., 415 U.S. 394 (1974); Fortnightly Corp. v. United Artists Tel., 392 U.S. 390 (1968).

[30]See Teleprompter Corp., 415 U.S. at 409-10; Fortnightly, 392 U.S. at 400.

[31]See Fortnightly, 392 U.S. at 399.

ceive and carry."[32] Cable systems were found to extend the viewing area and enlarge audience size.[33]

Subsequent provisions by Congress and the FCC specified protections and compensations for owners of copyrighted content in the original programming.[34] First, cable redelivery of television signals to local audiences was largely exempted from any form of copyright payment.[35] In this instance, Congress recognized that original audiences of such signals are not displaced if their transmission medium is changed from television antenna to cable. For such signals, copyright owners are fully compensated for their works through program fees paid by the broadcaster that maintains an intact viewing audience.

Cable operators who import signals to serve distant audiences must make payment to copyright owners who claim that their works were the subject of secondary transmissions.[36] Copyright owners in retransmitted programs now include movie studios, sports leagues, news providers, religious broadcasters, Canadian stations, and music claimants.[37] Compensation is established through compulsory licenses that are revised from time to time through Copyright Office hearings.[38] Compensation through compulsory royalties is reasonably instituted here to offset revenues that owners might have earned had their content been directly purchased.[39]

Except for the smallest cable systems, licensing fees for distant retransmissions are based on a specified percentage of the subscription and the advertising revenues earned by the cable operator; the appropriate percentage to be paid depends on the number of imported distant sig-

---

[32]*Id.* at 401.

[33]*See Teleprompter,* 415 U.S. at 412. The Court explained: By extending the range of viewability of a broadcast program, [cable] systems thus do not interfere in any traditional sense with the copyright holders' means of extracting recompense for their creativity or labor.... From the point of view of the copyright holders ... the compensation a broadcaster will be willing to pay for the use of copyrighted material will be calculated on the basis of the size of the direct broadcast market augmented by the size of the [cable] market.

[34]*See* 17 U.S.C. § 111(d)(1)(B) (2000).

[35]*See id.* § 111(b)–(c).

[36]*See id.* § 111(d)(3).

[37]*See, e.g.,* Ascertainment of Controversy for the 1998 Cable Royalty Funds, 65 Fed. Reg. 54,077, 54,078 (2000).

[38]*See* 17 U.S.C. § 111(d)(4).

[39]This action was similar to Congressional activity in 1909 that bestowed the first compulsory licenses for the reproduction of sheet music on piano rolls. Congress instituted in the 1909 Copyright Act a compulsory mechanical license for unauthorized reproductions of published sheet music on pianola rolls, which had earlier been cleared of infringement by a 1908 Supreme Court decision that found that the musical compositions on pianola rolls were not directly perceptible in the perforations themselves. *See* White-Smith Pub. Co. v. Apollo Co., 209 U.S. 1 (1908).

nals.[40] The pool of collected monies is paid to competing rights holders based on administrative rules that attempt to determine the relative worth of works.[41]

The FCC has more directly protected distant imports of programs from television networks and producers of syndicated content. It now proscribes distant imports that directly duplicate existing network or syndicated fare that are otherwise available through local broadcasters.[42] If permitted, retransmission would not only deny a licensing opportunity to program owners, but also harm the ratings and advertising revenues of local stations, which may otherwise have attracted the same viewers.

The general paradigm for cable retransmission may have reasonable applicability to the Internet regime. First, Internet retransmissions that largely preserve or enhance viewing audiences can be made exempt from copyright licensing and payments. Second, unlicensed retransmissions that may duplicate programs and displace viewers may pose considerable dangers to the broadcast model, and may require their complete proscription.

## MARKET FAILURE AND THE PUBLIC GOOD

Besides the possibility of market harm regarding the loss of unit sales and licensing revenues, we must consider two additional economic factors in the discussion.[43] First, because the transaction costs of licensing are economically prohibitive, certain limited uses of copyrighted material might be made freely transferable.

An economic justification for depriving a copyright owner of his market entitlement exists only when the possibility of consensual bargain has broken down in some way. Only where the desired transfer of resource use is unlikely to take place spontaneously, or where special circumstances such as market flaws impair the market's ordinary ability to serve as a measure of how resources should be allocated, is there an economic need for allowing nonconsensual transfer.[44]

---

[40]*See* 17 U.S.C. § 111(d)(1)(B).

[41]*See id.* (d)(1)(D)(4).

[42]Respective FCC rules regarding cable network non-duplication, syndicated exclusivity, and local sports blackout now appear at 47 C.F.R. § 76.92 (2001), 47 C.F.R. § 76.151 (2001), and 47 C.F.R. § 76.67 (2001). In implementing the Satellite Home Viewer Improvement Act, Congress directed the FCC to extend these rules appropriately to the satellite market. *See* 47 U.S.C. § 339(b)(1)(B) (Supp. I 2001). The FCC, on November 2, 2000, released a new Report and Order in this regard. *See In re Implementation of the Satellite Home Viewer Improvement Act of 1999: Application of Network Non-Duplication, Syndicated Exclusivity, and Sports Blackout Rules to Satellite Retransmissions of Broadcast Signals*, 15 F.C.C.R. 21 (Nov. 2, 2000).

[43]*See* Gordon, *supra* note 16, at 1614.

[44]*Id.* at 1615.

As a second related matter, copyright exemption of certain material is reasonable if the uncompensated transfer provides a social gain.

If market failure is present, the court should determine if the use is more valuable in the defendant's hands or in the hands of the copyright owner.... [F]air use is often found where defendant's use of the work is noncommercial and yields "external benefits," that is, benefits to society that go uncompensated. In the presence of such market failure, the price that the defendant user would offer for use of the work will often understate the real social value of his use. The courts in fair use cases frequently make intuitive estimates of social value.[45]

## TRANSACTIONS COSTS

Digital technology allows users to transform and combine broadcast material into new presentations. Combined applications may include the sequencing of two or more video clips, the simultaneous presentation of two copyrighted works (e.g., video and music), or the morphing of characters through digital techniques. Transformation can sometimes occur in an open source base of users who may make sequential adaptations of a work.[46] Licensing requirements in a number of these applications appear highly idiosyncratic to the specific needs of the presentation at hand.

Historically, licensing agencies confined themselves to individual and period-specific applications related to a single work, or a body of related works. For example, the American Society of Authors, Composers, and Publishers has licensed the right to make public performances of musical works in its catalog.[47] The Copyright Clearance Center has licensed the right to make photocopies of copyrighted texts,[48] and the Media Image Resource Alliance has licensed rights for photographs.[49] In the devolution of licensing contracts, businesses and public non-profit organizations (e.g., schools, libraries, religious organizations) were free to negotiate and contract for the right to use copyrighted material. These licenses were often blanket arrangements that allowed unconditional use of a work for a specified period of time.

---

[45]*Id.* at 1615–16.

[46]For a good collection of articles on the open source movement, see O'Reilly & Assoc., *at* http://www.oreilly.com/catalog/opensources/book/toc.html (last visited Jan. 24, 2002).

[47]*See* American Society of Composers, Authors and Publishers, *at* http://www.ascap.com (last visited Jan. 24, 2002).

[48]*See* Copyright Clearance Center, Inc., *at* http://www.copyright.com (last visited Jan. 24, 2002).

[49]*See* Media Image Resource Alliance, *at* http://www.mira.com (last visited Jan. 24, 2002) (warehousing stock photos).

It is not clear what kind of administrative domain will prevail for content used in multimedia or combinatorial presentations, where licensees will face the need to contract for the simultaneous use of a number of different works. For large and frequent users in businesses and non-profit public institutions, adaptive licensing mechanisms can be expected to result from the continued efforts and negotiations of related parties determined to spend the time necessary to make the system happen. Here, a constellation of rights organizations will evolve, including consortia, subscription agents, copyright collectives, rights clearance centers, and "one-stop shops."[50] These evolving institutions in intellectual property are the proper focus of the "new institutional economics," which suggests that facilitating market arrangements evolve as the clear need for them becomes recognized.[51]

For small uses, such as noncommercial applications by private associations of citizens, particular uses of copyrighted works may be repeated once or a small number of times. Per use licensing can be expected. It is not clear whether negotiations are practical, whether institutions will evolve, or whether the resulting licensing structure will be adaptive or efficient for such small uses. Transaction costs may be prohibitive to any small user if the appropriate licensing cannot be efficiently provided.[52]

Accordingly, if licensing were required for small uses, the associated costs might dissuade most efforts entirely. In the first place, a number of small users do not earn revenues for the content creators. Moreover, if the content involves multiple participants who simultaneously or sequentially edit works, the team would face the considerable task of assigning the licensing costs to all contributing participants.

---

[50]University consortia are teams of libraries that negotiate collectively on behalf of a group of individual members. Subscribing agents are commercial agents who negotiate usage contracts on behalf of one or many licensees. Copyright collectives negotiate contracts on behalf of their rights holders, such as in photo-reproduction or musical performances. Rights clearance centers grant licenses based on individual terms specified by the owner. "One-stop-shops" are a coalition of separate collective management organizations, which offer a centralized source for a number of related rights, such as photos and music, that would be particularly useful in multimedia production. *See* World Intellectual Property Organization, *at* http://www.wipo.org/aboutip/en/about_collective_mngt.html (last visited Jan. 24, 2001)

[51]*See* Robert P. Merges, *Contracting into Liability Rules: Intellectual Property Rights and Collective Rights Organizations*, 84 Cal. L. Rev. 1293, 1294 (1996).

[52]Transactions costs include drafting, negotiating, performance safeguarding, renegotiation, monitoring, and enforcement. *See* Oliver E. Williamson, The Economic Institutions of Capitalism 20–22 (Free Press 1985).

## THE PUBLIC GOOD

In awarding radio spectrum to television broadcasters, the government freely bestowed a substantial public asset that has considerably benefited stations, program producers, and advertisers.[53] The justification for such free takings, if any, lay in the capacity for broadcasters to disseminate vital public information, such as news, and provide hours of public interest programming.

Enhanced by personalization and user interactivity, Internet video may eventually enable, in both the U.S. and the world, a wider domain of news, historical, and cultural presentations. This may lead to (1) a wider "community of memory" with heightened historical awareness of important individuals and events;[54] (2) a "shared language" of words and images[55] that transcend all present modes of communication; (3) a heightened awareness of cultural diversity in an evolving communications network;[56] and (4) a character more capable of, and attentive to, actively engaging in the production and transformation of culture.[57] With an eye to provide and disperse information to the citizenry at large, Internet video may serve a considerable role in reaffirming public values, educating the citizenry, and informing healthy public debate.

In facilitating the dispersal of public information, Internet video may present common benefits that all citizens may share, and evidently has aspects of a non-excludable public good. From an economic perspective, free markets may underprovide such public goods as each consumer fails to internalize the gains that others may enjoy as the result of his activity. Collective action is often justified to correct for market failure when goods are public.

The need here for collective action would ideally implicate a social contract negotiated between the public representative (the government) and the private parties that participate in broadcasting and program production. Imagine a starting regime where competing television stations paid for radio spectrum in order to provide an audience base for their respective advertisers. In exchange for free access to the same radio spectrum, television broadcasters and content owners would agree

---

[53]See R. H. Coase, *The Federal Communications Commission*, 2 J.L. & Econ. 1 (1959).

[54]See Robert Neelly Bellah et al., Habits of the Heart 152-54 (Univ. of Cal. Press 1985).

[55]See Gerald Dworkin, *Moral Autonomy, in* Morals Science and Sociality, 156-61 (H. Tristam Engelhardt, Jr. & Daniel Callahan eds., 1978).

[56]See Richard B. Stewart, *Regulation in a Liberal State: The Role of Non-Commodity Values*, 92 Yale L.J. 1537, 1568-81 (1983).

[57]See William W. Fisher III, *Reconstructing the Fair Use Doctrine*, 101 Harv. L. Rev. 1661, 1768 (1988).

to exempt from copyright fees those retransmissions of their programs that serve a clear public purpose. For their part, retransmitters would agree to cede copyright exemptions if broadcasters could demonstrate that viewers of original programming were displaced. If this hypothetical resolution can be envisioned to appeal to all parties, the benefits of ex post collective action, and a mutually accommodative social contract, could be established.[58]

## FIRST ROUND EXAMPLES

Below is a list of four possible examples of how over-the-air signals that may reasonably be re-used are exempt from copyright law. These arrangements can be facilitated through voluntary negotiation or, more arguably, statute. They should not be taken as policy recommendations by this author. Rather, each example illustrates a preceding conceptual point from the text above, and is intended to stimulate thought and discussion.

### Video Clips

For non-commercial uses by online associations of private citizens, short video segments clipped from over-the-air broadcast programs might reasonably be exempted from copyright protection. For example, sports fans may assemble short clips of their favorite athletes, entertainment fans may be similarly attracted to their favorite performing artist, and study groups may use excerpts from religious or historical programs. To enable multimedia presentation, video clips might be sequenced, modified with new background music, or video "morphed."

With rights to make limited reuse of broadcast material, online communities of Internet users may combine and reformat material in an ongoing manner. The resulting video product may evolve from an open-source process that greatly enhances the democratic culture of the Internet, and draws on the creativity of its participants. Internet video will open content to new influences, expose people to new material, and greatly stimulate human thought and interaction.

Were the free reuse of video clips allowed for short applications, copyright owners could actually benefit from the process in a number of ways. Generally, a video clip of a program is not an appropriate substitute for the entire program from which it was derived. Non-commercial clipping would therefore not displace program audiences, and may actually advertise the show to new viewers previously unaware of its appeal.

---

[58]*See id.* at 1727.

Furthermore, a number of protective rules would be established in order to ensure that viewership of the underlying content is promoted.[59]
Clips would need to list the details of the original show in order to promote
viewership; required data would reasonably include name, local sta-
tion/network and viewing time of the original series. Takings in a secondary presentation could reasonably be limited in duration, number, and a
determined period of time after the broadcast in which they may be used.

Finally, content owners would retain the exclusive right to offer material
for commercial and public, non-profit uses. Commercial providers,
schools, and libraries can draw on popular fare to create more material,
which can be expected to increase the licensing revenues that they pay.
Viewers to fan club websites would presumably be more inspired to
hyperlink to commercial sites, to the benefit of the performer, the original
programmer, and possible advertisers. These hyperlinks would increase
traffic and commerce at no additional cost to the business.

## News Archiving

The presentation of archived news broadcasts to the population-at-large
would disperse important knowledge, increase historical awareness,
and enhance voter-based democracy. After a delay of a few days from an
original broadcast of a news event, it may be reasonable to permit free
retransmissions that may be edited, archived, and reformatted. With
rights to re-use news, packagers can enhance original content with related material and/or hyperlinks to other web sites. Independent commentators could then provide video with their own analysis. Key gains for
a democratic citizenry may appear in greater depth or diversity of opinions and the historical presentations that a broadcast newsroom would
not provide. Commercial applications here may be desirable, in order to
provide universities, institutions, and educated publications with the
monetary incentives to elevate the medium beyond present levels.

From the vantage of audiences, broadcast news is time-sensitive.
Therefore, it is unlikely that viewers will substitute between a current,
same-day news story and an archived version of the same news event
shown a few days later. If archiving were permitted, repackagers may
remove original advertisements, but must credit all original network
sources for borrowed material in order to promote the original program. To limit takings to just news clips, subsequent talk analysis, either
by anchormen or specialized talk programs, would not be eligible for

---

[59]The rules here should resemble existing statutory provisions now designed to ensure that compulsory licenses for non-interactive streams of sound recordings will establish protections that promote their sale. *See* 17 U.S.C. § 114(d)(2)(C) (2000).

free retransmission unless they were made the object of parody or direct criticism.[60]

## Local Time-Shifting

Since the Supreme Court's decision in Sony Corp. v. Universal City Studios, Inc.,[61] owners of videocassette recorders (VCRs) have had the right to capture television broadcasts for noncommercial use. While broadcast programs had advertising embedded in the original presentation, users had the manual capability to fast forward and bypass commercial messages. The importance of the matter now is greatly heightened, as new digital personalized video recorders (PVRs) are now available that provide automatic capabilities for bypassing commercials and distributing stripped programs over the Internet.[62]

The broadcasting industry could compete against ad-skipping and digital distribution by facilitating Internet-based retransmissions to provide time-shifting of local broadcast programs. Like cable retransmissions, time-shifting of local broadcast programming might qualify for a copyright exemption, provided existing advertisements are not displaced. The size of the viewing audience should be monitored and reported to the original station for the purposes of supporting its advertising ratings. To protect viewership of seasonal and syndicated reruns, retransmissions must be streamed (or downloaded with digital rights management for one protected viewing) and limited to a short subsequent period, such as one week, after the time of the original broadcast.

Internet-based, time-shifting services would offer consumers and broadcasters four key gains. First, with the requirement that commercials be preserved and audience size reported, television stations and programmers may find that Internet-based, time-shifting supports their advertising model better than PVRs or VCRs, which have neither obligation. Second, consumers may save space and avoid the costs of purchasing new equipment. Third, viewers can pay for time-shifting services on a subscription, or a per

---

[60]Campbell v. Acuff Rose Music, Inc., 510 U.S. 569, 575-76 (1994) (stating the rationale for exempting criticism and parody for copyright protection).

[61]464 U.S. 417 (1984).

[62]In November 2001, Sonicblue launched the ReplayTV 4000 digital video recorder, which will allow users to record programs onto a hard drive and pause live television. Moreover, consumers can skip commercials during playback and distribute programs to other ReplayTV 4000 owners via the Internet. On Oct. 31, ABC, CBS, NBC and their parent companies filed suit, alleging that the device allows consumers to make and distribute copyrighted programs without permission. The suit argues that such devices deprive the networks of revenue and reduce their incentive to produce new shows. *See* News.com, *Sonicblue to Launch DVR, Despite Suit, at* http://news.cnet.com/news/0-1006-200-8005769.html (Nov. 28, 2001).

unit basis, that allows greater flexibility in usage; downloads for more extended viewing can be made available for an additional payment. Finally, users will not have to preprogram the service; people may then have the opportunity to retrieve shows they may have forgotten or overlooked.

To protect against duplicate programming, signals could not be retransmitted beyond the local viewing region. Users would be required to enter zip codes, which would be checked against geographic information located on servers at the point-of-presence, where the Internet transport system interconnects with the local telephone exchange. Regionalization of signals can now be enforced with edge control agents that reside on peripheral servers and enable transaction validation, media encryption, and forensic embedding needed for accurate identification.[63]

Accurate audience measurement is also essential. MeasureCast's Streaming Audience Measurement Service now deploys software residing on a broadcaster's server and records its exact number of visits.[64] Data can be paired with demographic information that can be detailed from customer panel surveys. This is preferred to server log-file analysis, where data on servers can be manipulated by any party with access to the file, and where user reports can take up to three months to prepare.[65]

## Distant Program Imports

Like cable, Internet retransmitters might disseminate local television signals to distant audiences which otherwise might not be able to receive the program. Even without payment, the commercial gains to the original broadcasters here can be considerable. In a path-breaking business model, Ted Turner sold ad space during TV shows appearing on his local Atlanta station, WTBS, to national advertisers, who were willing to pay considerable amounts of money to reach the wider audience that distant retransmission enabled. After enjoying the benefits of free promotion of his advertisers for nearly twenty years, Turner further profited by converting his popular superstation channel to a cable channel in a sale to Time Warner.

In light of the Supreme Court's decisions in Teleprompter[66] and Fortnightly,[67] we can reasonably expect that Internet providers will be allowed to retransmit over-the-air television signals to distant broadcast regions without paying the original broadcaster. The issue remains whether pro-

---

[63]See Vidius, *A Service for the Control, Audit, and Protection of Online Media*, at http://www.vidius.com (last visited Dec. 7, 2000).

[64]See Measurecast, *An Analysis of Streaming Audience Measurement Methods*, at http://www.measurecast.com/docs/Audience_Measurement_Methods.pdf (last visited Oct. 22, 2000).

[65]See id.

[66]Teleprompter Corp. v. Columbia Broad. Sys., Inc., 415 U.S. 394 (1974).

[67]Fortnightly Corp. v. United Artists Tel., 392 U.S. 390 (1968).

gram producers will demand payment for re-use of copyrighted programs. While it is not appropriate to compel owners to give away material for distant imports, some producers might actually choose to grant retransmission authority for selected programs at considerably reduced, or even zero, rates. This is because retransmitters may have the capability to make localized and personalized measurements of audience tastes and characteristics, thereby providing an efficient means of establishing potential audience size in new markets for sales of programs to local cable operations.

## CONCLUSION

As more information is learned, initial categorizations may prove erroneous, and the borders that delineate rights and exemptions can then be suitably modified. Reversing Robert Merges' suggestion that exemptions be established after allowing markets some time to take shape, we then establish exemptions to "jump start" the process, but reserve the right to modify or vacate certain allowances if harm can later be demonstrated.[68]

Such a procedure would evidently be incrementalist and experimentalist: restricting considerations, limiting classifications, forsaking quantification, leaving options open, and allowing more information to come to the table in the end. A policy process that moves by incrementally changing specific rules is often preferable to wider hearings and rulemakings that may overtax available administrative channels for gathering information and judging outcomes. Forsaking quantitative measurement and a fully comprehensive menu of choices, we learn which outcomes provide satisfactory short-run results by purposely restricting decisions and limiting the necessary amounts of information. The resulting process is more procedurally, rather than economically, rational but can be sometimes compared favorably with policy that aims for a purported welfare-maximizing optimum.[69] Generally, such incrementalism is particularly applicable to policy-making in the open-ended world of digital technology and its ability to provide significant transformations of copyrighted content.

---

[68]*See generally* Robert P. Merges, *Intellectual Property and Costs of Commercial Exchange: A Review Essay*, 93 Mich. L. Rev. 1570 (1995).

[69]*See generally* C. E. Lindblom, *The Science of Muddling Through*, 19 Pub. Admin. Rev. 79 (1959). Lindblom compares *incrementalism* favorably with *rational comprehensive* policy that is elegant but often impractical; rational comprehensive policy tries to consider and weigh all factors, gather all relevant information, measure all relevant quantities, and willingly jump to extreme positions as logically justified. *See id.*

# Network Business Models
# and Strategies: The Role
# of Public Service Broadcasting

**Fritz Pleitgen**
*Westdeutscher Rundfunk* Köln

Samuel Goldwyn, the film producer, once said, "It's always difficult to make forecasts—especially about the future." Keeping his warning in mind, this chapter, nevertheless, attempts to look into the future.

A recent publicly discussed design for the future of the media can be found in the Bertelsmann Foundation's "Communications Order 2010." In this document, experts paint a pessimistic economic picture of the future, the "Narrowband Scenario." It is believed that the very high investment required has resulted in a hesitation in the development of the digital market. It is now up to the comprehensive program channels and the mass media. The optimistic "Broadband Scenario," on the other hand, assumes affordable prices for the use of the new media. The anticipated result is that in several years an interactive broadband device will take the role of today's television for domestic multimedia use. In addition, intelligent agents will presumably preselect and arrange a variety of content, according to individual consumer preferences, from a literally endless choice of offers. The role of broadcasting corporations would therefore be one of "content providers" among a multitude of others. Whereas one of these scenarios could be reassuring, the other should encourage working even more for the future.

In any case, it is difficult understand or subscribe to forecasts that foresee the end of broadcasting. Available data suggests that broadcasters will become agnostic about delivery technologies. Broadcasting and the Internet will be complementary rather than the latter leading to the detriment of the former. The extent to which viewers may want to use interactive facilities on their television screens may depend on their age and upbringing. However, even then, the television screen will not be used as a working tool. It is a device used in the home. By design and function it is intended to be part of the overall information and entertainment sphere that forms an integral part of private life, separate from public or working life. So far, only a few people want to surf the Internet on their TV screens. Interactive television offerings that add value to the services already expected from television are likely to be most successful.

There is a frequently posed question about the role of public broadcasting corporations in the 21st century. A story by Jorge Luis Borges, written at the beginning of the 1940s, provides some insight. The well-known Latin American author described the "Library of Babel." It is the ultimate library. Any book in any language is to be found in the countless bookshelves of the library; the library contains the collected knowledge of the present, the future, and the past. The librarians are infinitely proud of this universe of knowledge, which will answer all questions. But this only lasts until they discover the true character of the Library of Babel. The endless rows of books not only contain all the truths of the world, but also all the lies. For each claim, there are a thousand counterclaims and, even worse, what people are really looking for will never be found in the endless labyrinth of rooms, corridors, and shelves. The knowledge of the Library of Babel is virtual knowledge. It is impossible to get to the bottom of things, and the librarians are in deep despair about this realization.

The apparent promise of the Library of Babel—namely, to be able to answer all questions—is also reflected in today's cure for all ills, the "information society": This is a medial democracy where anyone should be able to have all relevant information at any time. Information is not knowledge. Information is only virtual knowledge. Information may generate knowledge, it can contribute to understanding, it can guide action, but only if one has free access to it, and especially, if one knows how to use it. For now, the human being in the digital world is an actor—less knowing, rather searching.

The mission of public broadcasting corporations in the digital age has not become obsolete, quite the contrary. In order to escape from this Babylonian aberration, the public broadcasting system, with its mission, would need to be invented if it had not already existed for many years. Public service broadcasters have an obligation to provide their audiences with a broad range of programs that suit the needs and expectations of the entire population, young and old, highly educated and less educated, fully active and disabled. There must be a broad and democratic dialogue between all members of society whether they belong to statistic majorities or minorities. Thus programs, or "content" as it may be called in the multimedia future, must be relevant. They need to reach civil society at large to make a difference. Because the way in which content is provided to the public changes, it must be changed in order to continue to fulfil this mission.

In the digital age, public service broadcasting holds the potential to become the communication platform for all. It cannot be manipulated as credibility is its success factor and independence is a precondition for credibility. Perception and awareness are basic preconditions for fulfilling this important mission. Thoroughly and self-critically, there has been much reflection on the nature of these basic preconditions:

- Do public broadcasters cannibalize themselves if they become too active on the Internet?
- Will Coca-Cola, IBM, and Comdirect become more important for public broadcasters as competitors than the old rivals?
- As an old medium, are public broadcasters in danger of being devoured by the new medium, even if they are not listed on the stock markets?
- Will public broadcasters enter into a spiral of permanent underfinancing because all their competitors are able to finance themselves and their services by e-commerce revenues, whereas they must forego classic offers for every effort on the web? Is the Internet thus devouring its public broadcasting children?
- As a content supplier, public broadcasters have the Internet's most valuable substance. Again and again, cooperation offers are made. What do responsible cooperations of the future look like?

There is hope for a renaissance of public service broadcasting, in times of an exploding and complex media range.

So, what is next? Certainly, whatever is done in the future, in keeping with past and present times, must correspond with the public service focus. Quality content that is relevant to society at large needs to reach audiences everywhere on any device, be it a stationary or portable television, computer, digital radio, or telephone.

Apocalyptic theses have accompanied the introduction of all new media since the introduction of writing. The Internet does not drive people away from television screens. On the contrary, in households with Internet access facilities, those networks that are strong on the Internet are also those tuned in more regularly on television and radio. At least that is the experience of ABC and NBC. This year's (2000) ARD/ZDF Online Study has also shown no overall decline in television use for Germany, from 1997 to 2000, although the use of the Internet is increasing.

As a public service broadcasting corporation, this convergence must be approached from two sides: the side of digitization and the side of the Internet. With public broadcasters' knowledge of technological developments and in fulfilment of their mission through the digitization of TV and radio programs, they are strategically placed at several levels.

Public broadcasters have been broadcasting their own digital ARD package on satellite, "ARD digital," since 1998. The electronic program guide (EPG) was the first feature to integrate text and audiovisual content on the same screen. Previously, the worlds of television programs and videotext were totally separate features watched alternatively on the television screen. With the next step, setting up the multimedia home platform vested with back channel opportunities, this interactive tool comes ever closer to its role as an orientation navigator.

Of course, this is only the beginning. The personal TV, technology such as TiVi Anytime, and other time shifting systems have the potential to completely personalize television schedules. This, in turn, will challenge the program scheduling policies of all broadcasters. Already today, some limited interactive services are offered as part of the digital package, and these possibilities will dramatically increase once broadband television networks have been digitized and upgraded with back channels.

The Internet is the natural and ideal partner of public service broadcasting. On radio and television, it already provides detailed reports, but the Internet will enable coverage of almost any aspect and interrelated dimension of a subject.

Through their various Web sites, the 10 regional independent ARD broadcasting organizations offer a multitude of content. Currently, they are in the process of interrelating their web activities. Thematic portals will be created offering services to the user and simplifying orientation on the web. A news portal will be started under the domain tagesschau.de. In addition, there are offerings from the areas of service, education, sport, and so on.

At WDR, online services have emerged as considerable content providers alongside television and radio. There has been implemented a portal here too, intended for the people of North Rhine-Westphalia. In the future, public service broadcasting corporations' online offers will not be separate from their traditional broadcast services. They will instead form an integral part of a program strategy providing quick updates on news and stock market headlines, sports scores, and the like during and in between regular programs.

No other mass medium mixes information, entertainment, and commercial advertising as consistently as the Internet. This is due to the genesis of this medium. From the very beginning, free content offers on the Internet were made possible only through advertising links. This interconnection is a major opportunity for public service broadcasting. These broadcasters do not want to empower the Internet user for the marketing or sale of specific products. They want to make users the sovereign users of the new medium.

In all suppositions made about users, their convenience, and their supposed use of their time budget, the joie de vivre and social life cannot be replaced by technical applications—no matter how clever.

# 12

# International Regulatory Issues

**Stephen Whittle**
*British Broadcasting Corporation*

The Internet has become the most talked about technological development of recent times. How is it possible that this "thing," which was developed as a device to decentralize knowledge and data in the event of nuclear attack, became a means of academic information exchange, was embraced by "techies" and finally business, and now excites so much passion or such miscalculation?

It is a timely moment to stop and reflect on where things are heading on the eve of what has been described as the fourth age of broadcasting. Even in the United States, home of the Internet and the World Wide Web, not everyone is convinced that change is positive. According to Max Frankel of the *New York Times,* "It is hard to avoid the conclusion that our remarkable, convulsive revolution in the technologies of communication has debased our standards of journalism and eroded our capacity for civil discourse. We are wallowing in information—but we are starved for understanding." How then is it possible to make sense of what is happening in the world of the Internet? After much consideration of the issues, it became apparent that regulation was likely neither the question nor the answer.

First, as with all revolutions, consider what is happening. There are some obvious opportunities and some equally obvious challenges. The positive side of what is happening is the enormous educational and democratic potential seen in what is made possible by the Internet. The information society offers the chance for a better informed citizen to make a real contribution to the national debate. The most obvious downside is that, as with other communications revolutions in the past, the gap between the

179

"haves" and the "have-nots," between the information rich and the information poor, widens both within and between societies. It is also possible
that the opportunity to enrich the many, by making information and power
available to all, will be subverted to make money for the few.

Should the mood be pessimistic? Or are there encouraging signs? This discussion starts as one that looks at the glass as half full rather than half empty.

## GLOBAL INEQUALITY

The growth in the Internet is well documented: Recent research reveals
that the Internet universe grew three times faster than television over the
same period of time (200 million Internet homes after 6 years, which is a
figure that took television 20 years). The growth is global. The United States
accounted for over 50% of all Internet homes by year 5, but there are already signs that the balance is shifting (see Table 12.1). New estimates
suggest that the fastest new growth will come in the Asia-Pacific region.
Discrepancies still exist in take-up.

The center of gravity is moving east and south and with it there is also a
remarkable growth in Internet languages, especially the Asian languages
and Spanish. Mexico, for example, sees more web use than the United
Kingdom, Germany, or France.

What the Internet underlines is the great gap that continues to divide
the world, not least the gap of age. But the Internet is not itself responsible
for inequalities in wealth, education, access to technology, infrastructure,
and so on. There are, of course, many side benefits. Look, for example, at

TABLE 12.1
*Internet Take-Up*

|                     | Millions | % of total |
|---------------------|----------|------------|
| World               | 201      | 100        |
| Africa              | 1.72     | 0.86       |
| Asia/Pacific        | 33.61    | 16.7       |
| Europe              | 47.15    | 23.5       |
| Canada/United States| 112.4    | 55.9       |
| Latin America       | 5.29     | 2.6        |

*Note:* Data from Nua Surveys, September 1999.

the comparative costs involved in sending material from Madagascar to the Ivory Coast by post, fax, and e-mail.

Also, thanks to UNESCO and others, Africa is being given start-up help to enable it to develop its own approach to the new electronic world. Identifying the problem is at least a start on the way to a solution. But again, it is important not to confuse the messenger with the message.

What is remarkable is how the e-world is being used to share knowledge, to make available expertise, and to break down divisions. For example, in Latin America, health care techniques and treatments are being shared so that they become more universally available. So, even in the midst of global inequality, there are some encouraging developments.

## CONTINENTAL DIFFERENTIATION

Even within the rich world, there are considerable differences in Internet adoption. In Europe, the Scandinavian countries have been very active in their promotion of the Internet. Finland has both the highest per capita usage of mobile phones and people with access to the Internet. However, some of the Latin countries are only just taking up the Internet in significant numbers (Pro Active International, 1999).

The Broadcasting Standards Commission in Britain noticed some of these differences in a recent study that looked at the way in which children used the screen. It found that in 1997 (Livingstone & Bovill, 1999), 7% of British 15- to 16-year-olds had access to Internet at home, as compared to 38% in Sweden. Part of the explanation lies in the fact that British parents are reluctant to allow their children out of the house when they are not attending school. So they provide "entertainment centers"—televisions, videos, music centers, games—in the bedroom, rather than learning zones full of books or personal computers.

## NATIONAL DIFFERENTIATION

Again in Britain, the differentials between upper and lower social grade households with regard to Internet access are more marked than in other countries such as the Netherlands or Scandinavia:

ABC1 14% (Upper- and middle-class households)

C2DE 2% (Working-class and low-income households)

The vast majority of children in Britain still only have Internet access at school. This is something the British government has noted and is taking steps to improve. The government is working to ensure that all schools are connected to the Internet, to improve the quality of teacher training, and is now also making computers available to low-income households. But there is still a long way to go. The study referred to earlier indicates that there is a further gap between children whose parents are information technology competent, and who can reinforce or even improve on school work, and those who are not competent.

## COMMUNICATIONS REVOLUTION?

All of this, of course, helps to put the Internet into perspective. It is not quite as all pervasive as its promoters would like people to believe. Indeed, it is probably quite unsuited to much that is promised. But, there is little doubt that it has revolutionized all kinds of communication. It is many different things at the same time: a cross between an information exchange, a library, a chat line, a shopping mall or banking hall, a post box or an entertainment center. It offers immense benefits and engenders numerous anxieties. It belongs to no one and no nation. Some see it as a great gift to freedom, others as an invitation to anarchy. It combines private and public functions in a unique way, but it is not lawless, and it does not present the same kind of issues that arise from the invasive potential of broadcasting.

Now a new wave of excitement is under way as broadband technology opens up the possibility of linking internet and television in new and more challenging ways, by blurring the obvious distinction that has existed up to now between a "pull" and a "push" medium. How much of this is real and how much is hype?

In a recent survey of e-entertainment, called, appropriately enough, "Thrills and Spills," *The Economist* was clearly skeptical. It noted the enormous sums of money being invested in e-entertainment by entertainment companies terrified of the challenge but excited by the prospects, because the Internet seems a way of delivering their goods directly at very little cost. It seems to make very targeted advertising possible while remaining cost effective.

But, as *The Economist* remarked, "the reality has not matched up to the vision." There seem to be two basic problems. One is the difficulty of distributing content on the Internet, the other is people's unwillingness to pay for anything beyond what they are already paying for Internet access. For example, music is easy to distribute, but it is hard to persuade people to pay for it. All that people seem prepared to pay for are *The Wall Street Journal,* some games, and a great deal of pornography.

If you look at Internet usage in the United States, the vast majority use it each week for e-mail (90%), search engines (70%), researching product purchases (44%), health (35%), and reading a newspaper (25%). The most visible entertainment use is game playing (22%). Internet virtues (e.g., freedom from censorship, speed, low distribution costs, global reach, and interactivity) seem to benefit the pornography business. It accounts for almost all paid content on the web. Almost one third of Americans now get news online at least once a week, although not all news sites are provided by newspapers, such as CNN, BBC, and others. Sport and niche businesses are also having some success.

It might be that broadband will make a difference, but so far the evidence shows that it is too slow taking off. It is still technically very complex to deliver, even via cable. The distribution problems have not encouraged the content industry. It is a vicious rather than virtuous circle and there is still the question of whether people will pay. Putting aside all the usual comments about Amazon.com or the crash in e-markets, very few sites succeed in charging customers for their wares. *The Wall Street Journal* is an honorable exception. The rate of e-advertising is also slowing. *Big Brother* in the United Kingdom offers another indicator. The most visited Web site in Britain attracted advertisers, but not sufficient to pay for the site, and virtually no e-commerce. The proportion of people clicking on through to the advertisements is falling to about 0.4%.

The difference, therefore, between television and the Internet remains stubbornly clear, despite the claims of Negroponte. To state the obvious, people use the television and the personal computer in quite different ways. Microsoft has not been able to turn the trick with Web TV. Digital television seems to hold more promise of satisfying consumers. So the conclusion has to be that the claims made for the Internet, even delivered via broadband, are unlikely to fundamentally change the world. It may prove to be the means of distribution for music. But books are likely to remain popular as books. Games, news, and sport have a web future. But moving pictures will remain elusive for some time to come.

## A NEED FOR REGULATION?

So what is the challenge of the Internet? Does it need regulating? Why? How? Clearly, context is key to any regulatory strategy. What is the nature of the service, the means of its access, the method of payment, the likely expectations of users, and so on?

A considerable proportion of what happens on the Internet clearly should excite little interest or concern. Private mail is and should remain private, from both employers and the state. Visitors to news sites, libraries, book shops, record stores, or shopping malls only attract attention in the real world when they are up to no good. The same should be true of the virtual environment. There are some issues here of consumer and data protection, that transactions have legal force, that fraud is no easier electronically than it is in the shopping mall, and that privacy rights are respected. British evidence suggests this is an area where government needs to do more to promote confidence. Clearly, too, there are issues of copyright protection, which require both technical sophistication as well as concerted international legal action.

The potential problems arise over the easier access for children and vulnerable people to material that might be considered either offensive or harmful. A balance, therefore, needs to be struck between consumer expectations, the protections necessary for commerce and creativity to flourish, for child protection, freedom of expression, and privacy rights.

The public policy and regulatory challenge is both how to strike that balance and how to enforce the judgment. Currently, the approaches taken vary from country to country and are based on cultural and historical traditions. Surveys of public opinion taken in Australia, Germany, Singapore, Britain, and the United States indicate quite different concerns. In Australia, sex is the issue. In Germany, it is race hate. But in Britain, it is a concern about the protection of financial data. These differences make it very hard to come up with a single approach to Internet content issues. But, it is equally important to be clear about the difference between illegal, unlawful, and harmful content.

The aspect of the Internet that usually excites most comment is child pornography. But that is also the issue of greatest consensus. There is no jurisdiction in the world that does not regard this as an illegal activity, regardless of the means of distribution. Here it is relatively easy to get consensus and joint action.

Unlawful content is more complicated, in part because what is unlawful in one place may not excite the same attention in another. States have very different attitudes toward everything from Nazi regalia to the limits of sexual expression, let alone the protection of copyright. Nevertheless, where such things are illegal, there is a legal remedy to pursue. It does not, nor should it, require an additional level of regulation.

Potentially harmful content is more difficult. First, the Internet probably would not be equated in any way with a broadcast medium because the viewer has to seek the material out. Second, again there is no simple definition of what should be considered harmful: sex sites, chat forums, or instructions on how to make a bomb. Rights of expression and defining what is harmful and to whom are also problematic.

Again, different places are offering different approaches. Singapore and Australia, for example, have chosen the route of direct regulation. The Australian Broadcasting Authority, for example, requires Internet service providers (ISPs) to issue codes of conduct, can consider complaints about sites judged to be hosting inappropriate or illegal content, and can issue takedown notices. In the first 6 months of this year, they received around 200 complaints and issued 60 plus takedown notices. The result was that most sites migrated offshore.

In other places, the approach is one of coregulation in which responsibilities are shared between government and industry, with the legislation providing the framework within which content and service providers operate by their own rules. There are signs that this approach is having more effect. The xxx.domain proposed by the pornography industry wants to keep a clear identity, and other providers are keen to indicate the kind of content people can expect to find. It is probably unrealistic to expect to deter 14-year-olds from at least sampling a sex site, but again the development of more sophisticated ratings and filtering systems by bodies like the Internet Content Rating Association (ICRA) is also helping to underpin parental responsibility. The ICRA reckons to have involved the 20,000 or so sites that account for at least 80% of the traffic. The e-world will never be entirely safe, like its real-life counterpart, but significant steps have been taken to offer protection to those who need it.

## WHAT ABOUT THE FUTURE?

Broadband does present positive opportunities, especially by providing choice, expanding horizons, and developing new forms of creative and commercial life. But, as already seen, it is far from clear what time scale is necessary or indeed what the likely drivers are going to be. The costs and the skills required suggest the obvious danger of new divides opening between the rich and the poor, as well as various transactional concerns.

The potential failure of entertainment content on the Internet could mean that broadband might tackle the high ground of education, culture, and democracy. Whether the Internet proves a viable means of delivery of the awesome potential for involvement, interactivity, and knowledge sharing, which was part of the original ambition of the Internet, remains a question. The challenge for public policy is to ensure, as with current terrestrial broadcast services, universal access at little or low cost, to a full range of public service and generalist services that impart educational, health, and employment information, as well as telephony's universal access, and interoperability.

There are big and difficult issues. Concepts dealing with must-carry provisions, ensuring diversity of voice and range of supply, as well as providing open access, are notions of public service and public interest that have found a fuller expression in public service broadcasting in Europe. The American approach has always been different and United States is still the lead Internet culture. It may be that the difficulties that the entertainment industry is having may be to the advantage of the public sector. After all, the resources made available via the Internet do lend themselves to distance learning, citizen participation, and dialogue between the government and the governed. A key challenge will be equipping people with the skills of media literacy. People need to develop the same critical judgment with the new media that they possess with the old, including how to read a text and discern its message and how to evaluate fact verses fiction, truth from falsehood, and so on.

The Internet offers the opportunity to underline the old freedoms of expression and of information that are vital to social, economic, cultural, and political development. These new technologies can and should be used to further everyone's rights to express, seek, receive, or impart information and ideas for the benefit of both the individual and society. The Internet offers the opportunity for a million or more flowers to bloom. Let's encourage the growth for everyone's mutual enrichment, and to enable an even greater participation in public life.

Any future regulatory framework needs to be based on the minimum statutory intervention necessary to safeguard the public interest, coupled with responsible self-regulation by content and service providers and empowered and confident users.

## SOME WAYS FORWARD

It is already clear that nothing stands still. The potential is obvious but so are the threats. There is still have time to act and encourage positive outcomes. First, governments should work to encourage, not stifle, the potential by opening up the education system as well as the very process of government and decision making itself. Every government department and public body should have a Web site that provides user-friendly information and access both to the decision-making process and the decision makers. Second, public access to the Internet should be made available at libraries or in other community centers. Morever, schools must be equipped with both technology and know-how. Third, the developments in digital broadcast technology could be used to make the resources available on the Internet cheap and easy to access. Fourth, cultural bodies could be encouraged to use their imagination and creativity in this new world and get online. Strategies must be developed to teach and support media literacy to empower citizens for the new world and place even greater emphasis on training both within the educational structure as well as for people who have left formal full-time education. Lastly, make the encouragement of the information society an objective of development agencies, both governmental and non-governmental, and open up adequate and low-cost networks for new services both within and between nations.

This revolution is here to stay. As one former revolutionary once put it: "The philosophers have analysed the world; the point however is to change it." The means exist. Do people have the will to achieve it?

## REFERENCES

Pro Active International (1999, November). Preliminary findings.
Livingstone, S., & Bovill, M. (1999). *Young People, New Media*. London School of Economics.

# Content and Culture

# 13

# Audience Demand for TV
# Over the Internet

John Carey
*Greystone Communications*

*The Wall Street Journal* declared fall 2000 to be the "Web's First Fall Season" (Wilde Mathews, 2000), referring to the large number of Web sites offering video streaming content. Is television on the web in the same position as broadcast television was in the late 1940s—ready to explode as an entertainment medium for millions of people—or is a more complex phenomenon about to unfold?

There are many pieces to this puzzle. The first is the rapid advance in technology and, at the same time, uncertainty about exactly what level of video service can be delivered over the web in what time frame (see Noll, chaps. 1 and 3 in this vol.). There is also a proliferation of terms such as parallel broadcasting, two-channel TV, webisodes, telewebbers and preloading, which emerge constantly and often fade into obscurity just as quickly. However, some of the new terms may presage significant changes in media usage patterns. Further, there is uncertainty concerning whether the personal computer and the television set will converge into one terminal or whether both technologies will continue as separate units, each with some added features.

There is no shortage of forecasts about what is going to happen and which organizations will win the eyeballs of web users in the next generation of services, even though the track record of such prognosticators is abysmal (Carey & Elton, 1996). Many argue that "content will be king" in the Web TV world and this has led dozens of groups to create new content or aggregate existing video content. However, timing is crucial. The field of web video content providers is already littered with failures or false starts (La Franco, 2000). Further, the cost of transmitting video through a broadband web pipe is relatively unexamined. Some preliminary analyses suggest that the cost in the near-term will be high.

At the same time, web video offers the potential to find new audiences for existing video content that currently reaches a limited market (e.g., independent documentary films) and to help recover some of the lost audience for television. That is, a number of research studies have indicated that people who use the web heavily say that they have cut back on TV viewing. If web users are shifting some of their media consumption to the web, then existing video suppliers might be able to recapture the loyalty of viewers by providing video programming in this new medium. However, it is unclear how much of the reported time displacement (i.e., watching the web instead of watching TV or doing other activities) is actually time stacking (i.e., doing multiple things at once).

This chapter focuses on one piece of the Web TV puzzle: Do audiences want to watch video on a personal computer and is there a demand for all of the associated features that could be provided, such as interactive television, customization of video, and two-way video telephone calls? It draws on some lessons from earlier technology trials, current research about web usage habits, and an in-home ethnographic study of people who have broadband web access.

## LESSONS FROM EARLIER NEW MEDIA

In the rush to market web video, there have been few attempts to learn lessons from past trials and market introductions of high bandwidth video over telephone wires and other new media. The value in such a review is not just to prevent a repetition of earlier mistakes. It is also to learn the positive lessons about how to introduce new media effectively. Further, in the case of earlier technology trials that failed, there are often key insights, which in hindsight might have prevented failure and can now be used with foresight to help achieve success during a reintroduction of the service.

A few specific lessons and some general learning from earlier trials are relevant. The first is to be cautious about technological gimmicks. For example, France Telecom has experimented with a new service that can create scents to accompany web content as well as television programs. A similar technology, "Smell-o-Vision," was introduced in movie theaters during the 1950s to help recapture audiences that had been spending more time at home watching television instead of going to the movies. However, consumers saw little benefit in adding smells to movies. Further, there was a significant technological problem: It was very difficult to get rid of a smell created for one scene in order to introduce a second smell for a subsequent scene.

There has also been discussion about creating two-way video phone calls over the web. Video telephony has a long history. In the past, problems included high cost, jerky pictures, and a general networking problem (i.e., there is little value in having a video telephone unless other people with whom you want to communicate also have a compatible video telephone). Even if these problems can be overcome, there is a more fundamental obstacle: Many people in earlier trials and marketplace services simply did not want to be seen (Noll & Woods, 1979).

Many current and planned video services on the web offer small samples of content. This has been problematic in the past. Consumers expect new services to match or exceed what they currently use. For example, when the Discovery Channel tested an on-demand video program service called *Your Choice TV,* it offered a small sample of programs in a few categories (e.g., a couple of soap operas and a few news programs from three or four cable and broadcast channels). Reactions were weak. Consumers expected the service to offer a robust variety of soap operas and news programs, not a limited sample.

Earlier introductions of new media also provide a lesson about the time it takes to develop or discover creative new applications for a technology. Indeed, as McLuhan pointed out (McLuhan, 1964), people tend to fill new media with content from earlier media. So, many early TV programs were radio programs adapted for TV and much of early radio programming consisted of vaudeville acts adapted for radio. It takes time to understand the characteristics of a new medium and create exciting content for it. Often, developers underestimate the time and cost to build creative content that captures audiences.

The past also teaches the value of simplicity. For most people, the anchor service on the web that they value most is not researching topics of interest or having access to millions of content sources. Rather, it is electronic mail—simple communication with individuals that people know or with whom they want to communicate. Much of the early video content on the web is offbeat (e.g., two popular films offered through the web are *Froggy in the Blender* and *Bikini Bandits*). This may reflect the taste of early adopter college students, but in historic terms it does not reflect the demand by mass audiences.

Perhaps the greatest challenge to web video is changing habits. Historically, people have changed viewing habits many times, but there is often a strong inertia factor. Web video faces the challenge that until recently people have used television sets to receive video and personal computers to receive text and graphics. More recently, some personal computers have provided DVD drives that can play movies and high-end personal computer (PC) games have included some video. Further, there is much more use of personal computers for entertainment compared to a decade ago. So, some of the building blocks for changing behavior may be in place.

## THE USER EXPERIENCE

In one sense, it is remarkable that video can be transmitted at all over the Internet. In 1996–1997, it was science fiction to believe that the Internet could handle video. Despite extraordinary advances in compression and streaming software, however, most user experiences with streaming video in late 2002 leaves much to be desired. Narrowband video streaming generally provides a poor user experience. Typically, the video is displayed in a small box on the screen. Further, the video may be choppy and the audio may be out of sync with the picture. Downloading video at narrowband access speeds can overcome the problem of choppy performance, but at narrowband speeds it may take several hours to download a 30-minute television program. Broadband access speeds improve performance, but most services still provide less than full screen video and often a user must wait for some of the video to be downloaded into cache. Broadband services at the edge of a network (i.e., on a server at a cable headend or a telephone company central office) improve performance considerably by avoiding traffic congestion in the backbone of the web or links between a video server and the backbone. Further, compression technology has improved significantly, allowing more video data to be squeezed into the bandwidth of a given channel. Compression algorithms are expected to continue improving during 2003 to 2005.

It may be argued that in the late 1940s and early 1950s, most television viewers watched video programs on small screens, often with snowy pictures and double images, or "ghosts," caused by the reflection of signals off buildings and other large structures. By analogy, it could be argued that web users will accept poor quality video because it is a unique new service over the web. However, early television in the home had little competition and even bad pictures were a technological marvel for the average person. TV over the Internet is one of several ways for households to access video.

What types of video content does a web user encounter? A review of early content models reveals that there are at least six categories with a mix of existing television or film content and some new variations. The first may be called sampling or providing a short excerpt from a longer television program or film, often to encourage a web user to watch the TV program on cable or purchase the film on a videocassette or DVD. A second and related category involves providing a promotional trailer for a TV program or film. Typically, these are the same trailers that appear on television, in movie theaters, or at the front of prerecorded videocassettes. Third, traditional third-party advertising appears on some Web sites (i.e., a video commercial for a soft drink company that appears on a general entertainment Web site). Curiously, to date there has been relatively little third-party video advertising on the web.

A fourth category of content includes full-length television programs and films. The latter category consists of short documentary or animation films and even full-length motion pictures. These are available as video streams or downloads. The fifth and sixth categories include new content models. These have been labeled two-channel TV and parallel broadcasting. Two-channel TV includes content on the web that complements regular TV programming. In this sense, the distribution of content for a program utilizes two channels, regular TV and the web. There are a few variations within this model. One variation involves text or rich media on a Web site that complements a TV program (e.g., a web game that allows a viewer of the TV program to play along in real time or asynchronously). In the United States, the History Channel offers an online game, *History IQ*, that accompanies a TV program and lets web users play against each other as well as against contestants on the TV program. Another variation involves streaming video that supplements the video content in a TV program, for example, some scenes that were not in the main channel broadcast. Two-channel TV fits within a large subset of user behavior that involves simultaneous use of the web and TV. Most of this simultaneous activity consists of unrelated TV and web usage—for example, a person who watches a sports program on TV while surfing the web for news content. By one estimate, nearly half of web users in the United States make some simultaneous use of the web and TV (Neel, 2000). This group has been dubbed "telewebbers."

Parallel broadcasting is the transmission of the same content on a cable or broadcast channel and on the web, at the same time or a similar time frame. Most examples of parallel broadcasting have been in Europe—for example, video coverage of the 2000 Summer Olympics in Sweden via both broadcast and webcast. However, there has been some activity in the United States as well (e.g., the parallel cablecasting and webcasting of MTV's *Direct Effect*. In a few cases, the webcast has included long and largely unedited footage of a TV program before it was televised (e.g., *Big Brother* in Europe and *Inside Cell Block F* on Court TV in the United States).

A review of the user experience with TV over the Internet also reveals that certain types of video work much better in a streaming web environment than other types. For example, video with a lot of motion does not work as well as video with less motion and few pans or zooms. Similarly, close-up shots are easier to view in small boxes on a screen than medium or long shots. Large text fonts are also more legible in this mini-TV environment and simple animation works well because it helps to mask choppy video. In general, the challenge is to adapt content to meet the characteristics of small screen, low frame rate TV.

## THE MARKETPLACE CONTEXT

There are at least four categories of video streaming and downloading service providers: original producers of video for the web, content aggregators who specialize in video streaming, video search engines and portals, and traditional video distributors such as cable or broadcast networks who offer some video on their Web sites. In reviewing the types of content and services offered, it is important to note first that video streaming and downloading activities in Europe are at least equal to if not greater than in the United States. Overall, there were at least 50 trials of TV over broadband in Europe and the United States and many additional services for the universe of broadband and narrowband users. Several groups have developed robust video streaming services. Further, a few groups have begun or plan to charge fees for their video streaming content.

However, a much greater number of existing Web sites have been "testing the waters" with video streaming or downloading by offering samples of video content and occasional special events such as live sports coverage. Nearly all of the broadcast and cable networks in the United States offer samples of video streaming content on their Web sites. There has also been a great deal of experimentation with video streaming formats and content. For example, in the United States, some broadband ISPs have let consumers upload video clips like the video of a family wedding, store it on their server, and make it accessible through video streaming to other family members or the general public. In Germany, two television stations have created original soap operas for the web. RTL has created *Zwischen den Stunden* (*Between the Hours*), a 3½-minute-long soap opera that is webcast three times per week. Germany's public broadcasting network, ZDF, has created another soap opera, *Etage Zwo* (*Second Floor*). In the United Kingdom, British Telecom has been active in developing video services for its broadband customers.

## AN ETHNOGRAPHIC STUDY OF BROADBAND USERS

If video over the web is to be viable, then it will almost certainly require broadband access by users. Many issues then follow: What quality of video content can broadband provide; how fast will broadband services rollout in the marketplace; how quickly will consumer users adopt it; and how will people use the broadband web once they have it? The last question was addressed in an ethnographic study of broadband web households that is reported here.

Ethnography is a research methodology that was developed in anthropology for the study of distant cultures. Anthropologists would live with a native group over a period of months or years and write-up a detailed description of the culture based on observations and interviews. Ethnography was later adapted for the study of Western cultures and the behavior of people in everyday life (e.g., in the work of Erving Goffman) (Goffman, 1959). In the last decade, a number of researchers have utilized this technique for the study of new media use and effects, both for well-established media (e.g., television) and new media (e.g., interactive television) (Carey, 1996; Moores, 1996; Silverstone, 1994).

The study involved in-depth interviews with broadband users (all had cable modem access) in their homes and observations of how they interacted with web content as well as other household media. This small, qualitative study included 18 people in 12 households located in the northeast United States (New Jersey, Massachusetts, and New Hampshire). This form of research complements larger sample, quantitative studies. It is particularly suited to discovering new patterns of behavior and generating hypotheses that can then be tested in surveys or large-scale audience measurement research.

The people in this study had broadband service for periods ranging from 3 months to 2 years; the average was just over 1 year. Approximately one half were classic early adopters (Rogers, 1995). They wanted to be the first person in their neighborhood to have broadband service and some of them worked in computer-related professions. However, an equal number had only moderate interest in technology and did not have homes filled with electronic gadgets. One was a music teacher, another was a professional fisherman, and a third was a salesman.

The group in the study adopted broadband for a variety of reasons. Some recognized that it was the "latest and greatest" way to access the web. They actively sought out service providers in order to be the first to get high-speed web access. Others adopted cable modem service because their single telephone line was being tied up by Internet usage and this seemed to be a better alternative than getting a second phone line. And for some, broadband was adopted as the first web access service in their homes. A number of households were also influenced by the experience of high-speed web access at work or school. Most of the households were price sensitive, but they did a "back of the envelope" calculation and determined that cable modem service cost no more than dial-up service, once the additional telephone charges associated with a dial-up Internet service provider (ISP) were calculated.

It is important to note that no one in these homes used the term "broadband" or adopted cable modem service in order to receive services (e.g., video) that require broadband for an acceptable user experience. Rather, they talked about "high-speed" web access and adopted the service simply to get faster connections to the regular web content. Most understood the concept of broadband service, but it was not part of their everyday lexicon.

## Location of Computers

In the homes visited, computer(s) were located in a number of different rooms, including bedrooms, living rooms, and dens. In three of the households, a new type of room emerged—the computer room. To an outside visitor, these computer rooms appeared to be a spare bedroom or a den, but household members identified the room by name (they called it, the "computer room") and by association with a computer that defined the space and how it was used. One person, who had recently purchased a house, said that when he was looking for a new home it was important that a potential house have a space that could become the computer room. This behavior is much the same as when people select a new house based on kitchen size or the layout of living room and dining room spaces.

However, household space is a much more complex phenomenon than the simple labeling of rooms such as kitchen, den, living room, or even computer room. One important dimension is integration or openness versus seclusion. That is, some rooms are isolated from the rest of the house in terms of traffic patterns and used primarily by one person, whereas others are open to traffic and use by multiple family members. The computer(s) in these broadband household were located in both open and secluded areas, but the usage patterns were quite different based on their location. In secluded areas, they were a "cave for the hacker" and used primarily by one person. Also, the computer defined many of these rooms and other objects in the room were situated to support the central focus of the room—the computer. In open areas, the PC was used by multiple family members and the computer was integrated within the room rather than a way of defining it.

The function of a room can change by time of day or the positioning of objects within the room. These dimensions in turn can affect the function of a computer in the room and how people use the web. For example, some households in the study included people who work at home. They used the computer for their work during the day in a home office, but after work hours the room's function changed to a den and the computer was used for entertainment. Others reported that they too visited different Web sites based on time of day. Television usage also varies by time of day. However, most television programming, which is scheduled, changes by time of day to meet the interests of different user groups who are watching at that time as well as the changing interests of audiences over the course of a day. Relatively little web content is scheduled and therefore does not change much based on time of day. Nonetheless, the functionality of the web for many users changes by time of day.

In one household, the computer was located in a bedroom and was enclosed within an armoire that had two glass doors with curtains. The primary broadband user was a male, whose wife did not like technology in the bedroom. When the computer was not in use, the curtained doors to the armoire effectively closed off the computer from the bedroom.

A number of the computers in the broadband households were decorated with stuffed animals on top of the monitor or next to it. Some people placed family photographs and other memorabilia next to the monitor. These forms of decoration are probably not related specifically to broadband, but to a longer term trend of accepting the computer as a social and entertainment object within the household and not just a work tool. This treatment of the computer as a person or friendly object has been documented by Reeves and others in a variety of settings (Reeves & Nass, 1996). It is significant because it mirrors the treatment of televisions within households in the 1960s and 1970s, when family photographs, trophies, and other personal memorabilia often adorned the top of the TV set. These personal objects were later moved to the side of the TV and replaced by cable boxes, VCRs, or videogame consoles that were connected to the TV.

## Colocation with Other PCs and TVs

Nearly all of the households in this study had multiple PCs (some had three or four PCs, including laptops); one in three of the households had two PCs linked together in a home network. More remarkable, many homes had two PCs in the same room. In some cases, this facilitated the home network, but in other cases they were not linked. Rather, the two PCs in the same room indicated an intensive use of PCs by multiple family members in a room that was defined by the presence of the computers. This is not to suggest that they were all work spaces like multiple cubicles in an office setting. Typically, they were assigned to different family members who shared the space and used them for multiple purposes. This is another important dimension to computer use in homes. As households accumulate multiple PCs, some are shared and some are used primarily or exclusively by one person. TV sets in multiple TV households (a common pattern in the United States) also share this characteristic: Some TVs are used by multiple family members and some are "personal" TVs used primarily by one person.

In addition, people in households with multiple PCs made a clear distinction between online and offline computers. Typically, offline computers were older legacy systems assigned to small children for game playing or to household members with less need for or interest in computers. Online computers were newer and had a higher status by virtue of their faster speed and access to the web.

In three quarters of the homes, there was a TV in the same room as the PC or an adjoining space (e.g., in many U.S. homes, the living room and dining room are both part of one shared space). The distances from the PC monitor to the TV varied from 4 feet to 15 feet; most were from 5 to 8 feet apart. PCs and TVs were frequently on at the same time in these households. The orientation of the PC monitor and the TV set was as important as the distance between them in determining how or if the two were used together (shared TV and PC uses are discussed later in this chapter). In some cases, a person at the PC could glance slightly to the left or right and see the TV screen, whereas in other cases the TV was completely behind a person seated at the PC.

## The Changing Behavior of Broadband Web Users

In the households studied during this research, there were many new ways that people are using the web and many changes from earlier patterns of web use. The underlying appetites for web content and services have not changed dramatically (i.e., people still want information, entertainment, shopping services, and communications with others), but the ways in which these appetites are satisfied have changed and many other behaviors have emerged based on the characteristics of broadband web access.

Perhaps the most significant characteristic of the broadband web is not high speed but the fact that it is always connected. In most of the broadband households studied, the computer was on and connected to the web whenever anyone was in the house, much like TVs in many U.S. households (Bouvard & Kurtzman, 2000, p. 1). When the web is always on, it is possible to get e-mail throughout the day, to go to the web quickly for simple information such as a weather report or to check movie listings, and to use the web for background entertainment while doing something else (e.g., playing radio from the web while reading). All of these activities were common in the households studied.

When people get e-mail throughout the day, it starts to take on some of the functionality of a telephone. First, in order to know that new e-mail was coming in, many households set a tone to ring or an artificial voice to speak whenever the e-mail arrived. In this way, they could hear the tone or voice even if they were in another part of the house, much like a telephone ringing. Second, the constant availability of e-mail encouraged some to develop a relationship with others that relied on near real-time communication. That is, others learned over time that these people in broadband households would get messages almost instantly and could reply very quickly. For example, a professional fisherman in one broadband household used e-mail to schedule his work. This required a quick back and forth negotiation that formerly was done through telephone calls. Indeed, some people in these broadband households used the word "talk" when referring to their e-mail exchanges with others. For example, one woman said, "I talk to my dad in Seattle" when referring to her regular e-mail exchanges with her father. In addition, the constant reliance on e-mail led one couple to use it as a replacement for household notes. Previously, if one of them went out to the store, that person would leave a note on the refrigerator door for the other. They now used e-mail, knowing that the other would check e-mail on entry into the house, much as they used to check for notes on the refrigerator door. This constant use of e-mail was not without problems. A few people reported that junk e-mail was now a greater hassle because it got their attention right away; at least one person had turned off the e-mail alert sound as a result.

The "always on" feature of the broadband connection, along with the more pleasant experience of high speed access, encouraged many to spend much more time on the web—approaching the time many spend with TV. Other research supports these reports of greater time on the web with broadband access (Bouvard & Kurtzman, 2000, p. 3). Many said they were on the web for 3 or 4 hours per day, including brief sessions throughout the day and extended sessions at a few points during the day or evening. One person who worked at home (he was a day trader of stocks) said that he was on the web 14 hours per day. Spending more time on the web was not necessarily a positive experience for everyone. A few people indicated that they were concerned about managing their time and the web seemed to make great demands on their time.

### Impacts on Web Navigation and Features

The characteristics of broadband web access along with the evolving behaviors of broadband web users have led to several changes in how people in these households navigate the web and use certain types of sites. The first and most obvious observation in watching these people use the web is that many no longer have a home page and all rely less on home pages compared to dial-up web users. Less reliance on home pages is related to the always on condition of broadband web when people are at home. Unlike dial-up, where you start at a home page each time you access the web, the broadband web user finds the computer sitting at the site where it was left during the last session. In this sense, it is like television that remains on the last channel watched when it is turned on. However, some of the broadband users went beyond this and eliminated a home page completely. When they turn on their PC and click on a browser, it displays an empty screen. Others, especially those with less web experience, did use the home page of their broadband service provider (BSP). They found the home page a useful starting point for sessions, especially for accessing e-mail and local weather. Collectively, these patterns suggest that the future of BSP home pages may be challenged as broadband becomes more commonplace. At a minimum, the home page will have to compete for the attention of users by providing needed and wanted services.

Other observed navigation patterns and use of web features may be related to the experience of the web users in this study as much as broadband access. These patterns included the use of multiple browsers (many recognized that one browser performed better at certain sites and another browser performed better at other sites); less searching than they did in the past (they have found favorite sites and tend to stay within them); a frustration with bookmarks that have grown over time, often to 100 or more, and become unwieldy (some abandoned bookmarks, others organized them into categories, and a few cut back sharply on the number of bookmarks); and the formation of clear paths across sites that they use regularly (often there is more than one path based on time of day or functional needs, e.g., monitoring stocks). They did use some portal sites, but not so much for searching as for content. For example, many used Yahoo! for financial information, shopping, and games. It served more as a mall than a portal. Some of the more experienced web users opened four or five sites at a time, then reduced them. A web session consisted of circling round the four or five sites, opening and closing them. One person had two monitors in order to display two sites at a time. In addition, most of the web users in the study had a relatively small number of anchor sites that they would visit multiple times during the day. In some ways, these patterns of web use resemble radio usage, where people have a few preset favorites that they listen to regularly. With these web users, however, their usage behavior included a combination of regular visits to a small number of anchor sites plus occasional searches outside those anchor sites.

## Integrating Broadband Web Usage Within Family Patterns

Usage of broadband web services is shaped by existing family patterns and, in turn, it influences some everyday family behavior patterns. One notable observation was that some children used web services, especially broadband games, as a group. Two children would sit together at the PC and play a game together or alternate turns. Further, they would talk about the games—for example, giving tips about how to get past an obstacle. The broadband service was also used as a babysitter by some parents, who sat a child down in front of a game or other activity while the parent did household chores. Although adults did not share the broadband computer at the same time, some adults sat in the same room where there was a PC and a TV, talking about what each was watching. In addition, adults reported that parental controls were very important for controlling web access by their children, but no one actually used the parental controls that were available on the broadband service.

The location of the broadband PC or PCs appeared to follow from existing family patterns of communication and previous use of PCs. That is, in some cases the broadband PC was in an open family area and was used by multiple family members whereas in other cases it was in a secluded area and used primarily by one person. Although existing family patterns may have influenced where the broadband service would be located, the heavy use of broadband may be strengthening existing patterns of social integration or isolation. The hacker in his cave is now spending more time away from the family and those who integrated broadband among multiple family members are now using broadband for intra-family communication, sharing broadband content and, in some cases where a PC and TV are colocated, spending more time together. That is, some people reported that previously they often split up to watch separate TV programs in separate rooms, but now they spend more time together in the same space with one using the broadband PC and the other using TV.

## Multitasking with Other Media

The effects of the web on the use of TV and other media are complex (DiMaggio, Hargittai, Neuman, & Robinson, 2001). Nearly everyone in the study reported here said that they were using television less as they increased their web usage. This is probably correct. However, much observed behavior and discussions with them about the details of media usage suggested that there is a lot of multitasking or consumption of the broadband web along with TV, radio, and other media. Further, there is some multitasking within the broadband web itself (e.g., playing a web radio station in the background while exploring other sites on screen or watching a web video in a small frame on a Web site while taking in other content on the same screen).

Based on observations within homes and discussions with broadband web users, there are at least eight ways that the web and TV are used together:

1. The TV is used as a background sound while surfing the web.
2. A person alternates in small or large blocks of time between watching TV and surfing the web (here, the TV and PC monitor can be located anywhere in the room, as long as the person can move his head or swivel a chair to view either one). For example, a person watches TV while waiting for a Web site to load, then uses the TV as background until something catches his ear and he turns to the TV again.
3. A person simultaneously takes in TV and the web (here, the TV and PC monitor must be within the peripheral vision of the user). For example, a person watches a sporting event on TV and surfs entertainment sites simultaneously.
4. A person watches TV and waits for e-mail to arrive.
5. Different people in the same room are watching TV or surfing the web.
6. A person chats online about a TV program that is currently on and that he or she is watching.
7. A person watches a TV program or channel and simultaneously visits the Web site for that TV program or channel.
8. A person goes to the Web site of a TV program or channel at some point after watching the TV program or channel.

There are a number of variations to these patterns. For example, in some households, there were two PCs and a TV in the same room, with complex interactions among users and media. Also, in one household, there were two TVs and two PCs in the same space (an apartment with an open floor plan for the kitchen, dining room, and living room). It should also be noted that there was a high awareness of Web sites for TV channels among people in these households.

Survey research has indicated that there is more streaming and downloading of audio than video on the web (Bouvard & Kurtzman, 2000, p. 10). This was the case in the broadband homes that were visited for this study. There were four observed patterns of multitasking with audio: listening to over-the-air radio while surfing the web (this was reported to be declining), listening to a web radio station in background while surfing the web (this was reported to be increasing), listening to audio files such as news clips or music while surfing other web content, and downloading MP3 files in background to a recorder while surfing the web. Much of the multitasking with audio was completely within the web. In addition, a few people used the web to time shift radio programs (e.g., listening to previous episodes of *Prairie Home Companion* at the National Public Radio Web site).

## TV Over the Web

No one in the households that were part of this research adopted broadband to get television over the web. They adopted it to get faster access to regular web content. However, many discovered video content over time and began to use it. They accessed video news clips, sports clips, and some short films. They also valued the ability to get video from TV stations in other markets (where they lived previously) and they looked to the web for video when there was a breaking news story as well as for news activities that were scheduled (e.g., a space shuttle launch that was scheduled for a specific time). In addition, some believed that video of a breaking news event was likely to be placed on the web before it was telecast.

All of this video viewing on the web involved relatively short clips. They did not watch any full-length motion pictures or television programs. In addition, a few households had sent short video clips of family activities (recorded on a camcorder) to other family members in distant cities. Reactions to video on the web were mixed. Some people liked the video clips a lot, even though they were in small boxes on the screen. These people tended to be younger and spent the greatest amount of time on the web. Reactions to sending video clips were also positive among those who had tried it. Older users and people with high-end home theater TVs were less enthusiastic about video over the web. They wanted to see television that looked like television, not a small box within a web screen. In addition, some people pointed out the limitations of older PCs in accessing video over the web, even through a broadband connection. Older PCs lack the processing power to handle video and they do not have the storage space for large video files.

## CONCLUSIONS

This review of broadband users suggests that there is a latent appetite for video delivered over the web based on the evolving behavior of broadband web users. However, in order for this to become active demand by a mass audience, web video will have to meet a higher standard than is currently delivered under most conditions. This higher standard includes full screen video, a frame rate that approaches regular television, and a delivery method that is close to real streaming as opposed to long downloads. To achieve this, streaming video will most likely have to be delivered from servers at the edge of a network in cable headends, satellite network operating centers (NOCs), and DSL central offices, and it may require advances in compression algorithms. The work required to achieve this should not be underestimated and expectations should not be set too high, as has been done often in the past. Setting unrealistic expectations can lead to judgments of failure for a technology when in fact the technology simply needed more time to develop. Along with this, it will be important for program distributors and other video content providers to not penalize end-users with complex access procedures or added hardware in order to protect copyrighted material and manage digital rights. Such concerns and the resulting roadblocks that were placed in front of consumers delayed the widespread use of videocassettes, CD-ROM software, and video-on-demand (VOD), among other technologies.

It is also important to recognize where the current broadband market is located and how people are using it, as well as how the broadband web is evolving. Many broadband users are located in work environments and many are located in universities, although the home broadband market is growing at a rapid pace. Some groups (e.g., Yahoo! Finance Vision) have developed video services for the workplace and others are aiming at the growing home environment of broadband users. Further, most current broadband users adopt the service for high-speed access to regular Web sites, not video. Many discover video once the service is in place, but broadband users in general do not yet have a high awareness of web video. Indeed, if current broadband users tried web video when they had narrowband dial-up access to the web, they were probably disappointed and may not actively seek it now. At the same time, web usage behavior has changed for many of those in a broadband environment: People spend more time on the web; it is likely to be always on when they are at home; a number of people use it heavily for entertainment; many use the broadband web with television to complement and enhance the television experience; and there is some group consumption of the web by children. These patterns of usage support the evolving use of the web for video entertainment once it is a robust experience.

What about interactivity, customization, and other features that broadband video could provide? Do people want these added features? There is no simple answer. Some video-related services that have fared poorly in the past (e.g., video telephones) will face the same obstacles in the new broadband environment, such as concerns about answering a video telephone when people are in their underwear. However, some new variations of video telephones may prove to be popular (e.g., Fox Sports and MTV in the United States are experimenting with quiz programs in which people at home can participate in the TV game show by transmitting voice and images over the web via webcams).

Interactive television has been tried many times before, with mixed results (e.g., Warner-Amex's Qube system in Columbus, Ohio, and Time Warner's Full Service Network in Orlando, Florida) (Carey, 1996). However, the concept of interactive television has evolved to include a broad spectrum of applications with only moderate interaction (the concern in the past was that television usage was relatively passive and people would not want to press a lot of buttons while watching TV) such as video-on-demand and interactive TV program guides. Further, the web environment is inherently interactive. People interact with content all the time. So, the web may be a more benign environment to test various forms of interactive television. Indeed, some have suggested that the web could serve as a bridge from passive television to interactive television because it is an interactive medium (Cairncross, 1997). The same argument may be advanced for customization and personalization of video content. Customization and personalization of web content have been popular. Expectations for these features are likely to carry over to web video.

Audience behavior in a broadband environment can also inform discussions about convergence. To some degree, there has been a technological convergence of the personal computer and the television set (Forman & Saint John, 2000). However, there is no evidence that in the near term audiences will abandon one medium for the other. Rather, it appears that televisions are adding some PC features, computers are adding some video features, and the two will compete for the time and attention of audiences seeking entertainment. However, there may be as much parallel activity and complementarity as competition in the near term.

Discussions about advances in web technology, dealmaking by program distributors, and marketplace analysis of consumer behavior should not obscure the need for creativity and discovery of new content models for web television. In the short term, web video will borrow content models from television and film. However, the broadband web is an evolving, multidimensional space of text, rich multimedia, audio, and video with changing navigation and usage patterns. Over the longer term, creative artists must be given the opportunity to explore new program models that flow from an understanding of this multidimensional space and to build exciting new content that has never existed before.

The development of video over the web is in a very early stage. As with television in the late 1940s, nobody knows where it is headed. What impacts will a next-generation web have on individuals, families, business, politics, and society? It is not too early to begin asking questions and to set an agenda of research topics to explore. Hopefully, this series has contributed to setting such an agenda and this chapter has contributed to an understanding of one piece of the puzzle: audience behavior.

## REFERENCES

Bouvard, P., & Kurtzman, W. (2000). *The Broadband Revolution: How Superfast Internet Access Changes Media Habits in American Households.* New York: The Arbitron Company and Coleman.

Carey, J. (1996). *An Ethnographic Study of Interactive Television.* Edinburgh, Scotland: University of Edinburgh UnivEd.

Carey, J., & Elton, M. C. J. (1996). Forecasting the demand for new consumer services: challenges and alternatives. In R. R. Dholakia, N. Mundorf, & N. Dholakia (Eds.). *New Infotainment Technologies in the home: Demand-Side Perspectives,* (pp. 35–57). Mahwah, NJ: Lawrence Erlbaum Associates.

DiMaggio, P., Hargittai, E., Neuman, W. R., & Robinson, J. (2001). The internet's impact on society, *Annual Review of Sociology.*

Goffman, E. (1959). *The Presentation of Self in Everyday Life.* New York: Anchor Books.

La Franco, R. (2000, November 13). Hollywood's funk. *Red Herring,* pp. 93–98.

Marriott, M. (2000, September 28). Merging TV with the internet. *New York Times,* p. G10.

McLuhan, M. (1964). *Understanding Media: The Extensions of Man.* New York: McGraw-Hill.

Moores, S. (1996). *Satellite Television in Everyday Life.* Luton, UK: John Libbey Media.

Mowrey, M. (2000, October 2). Streaming bleeds cash. *The Industry Standard,* p. 173.

Neel, K. C. (2000, October 9). Starz! Targets Web Viewers. *Cable World,* p. 28.

Noll, A. M. (1999). The evolution of television technology. In D. Gerbarg (Ed.), *The Economic, Technology and Content of Digital TV* (p. 11). Norwell, MA: Kluwer.

Noll, M., & Woods, J. (1979, March). The use of a picturephone in a hospital. *Telecommunications Policy,* pp. 29–36.

Reeves, B., & Nass, C. (1996). *The Media Equation: How People Treat Computers, Television and the new Media Like Real People and Places.* Cambridge, UK: Cambridge University Press.

Reynolds, M. (2000, August 14). TV targets the Internet. *Cable World,* pp. 18–22.

Robinson, J., & Godbey, G. (1997). *Time for Life: The Surprising Ways Americans Use Their Time.* University Park, PA: The Pennsylvania State University Press.

Rogers, E. (1995). *Diffusion of Innovations* (4th ed.). New York: The Free Press.

Silverstone, R. (1994). *Television and Everyday Life.* London: Routledge.

Wilde Mathews, A. (2000, September 5). The web's first fall season. *The Wall Street Journal.* p. B1.

# 14

# Content Models:
# Will IPTV Be More of the Same,
# or Different?

**Jeffrey Hart**
*Indiana University*

Will the content of Internet protocol television (IPTV) be different from that provided by the traditional video delivery systems of broadcast television, video recordings, cable television, and satellite television? How are the new IPTV businesses structured? What new forms of intermediation are replacing older forms in this market? Are things moving away from a mass audience model for high-end television toward a more niche-oriented approach? These are some questions that are addressed in this chapter. Before directly tackling these issues, however, some background information about where things stand in this highly dynamic market is offered. Remarks are limited almost entirely to events occurring in the United States, even though there is some significant activity in other countries.

## THE SPREAD OF BROADBAND ACCESS

Many businesses and educational institutions already have high-speed Internet access via the purchase or leasing of T-1/T-3 lines or faster optical fiber networks. Typically, the individual user on a business network is connected via an ethernet connection (10–100 Mbps) to the enterprise network that is itself connected at a higher rate of transfer. Home and small business users are increasingly getting comparable access via digital sub-

205

scriber line (DSL) and cable modem connections. In 1999, according to Forrester Research, approximately 2.6 million households with personal computers (PCs) in the United States possessed a broadband connection out of the 44.8 million households with Internet access. The same organization projected a growth in broadband PC access to 37.5 million households by 2004 out of a total 80 million households online (Schwartz with Bernoff & Dorsey, 1998).

Cable modems have been marketed faster and in more areas than DSL connections. The cable operators have been more aggressive than the telephone companies in marketing cable modem services. Nevertheless, it is projected that the number of DSL households will soon exceed the number of cable modem households as the number of subscribers to both types of access continues to increase rapidly.

An additional 2.8 million households had access to broadband digital video via digital set-top boxes (STBs) in 1999. STBs give users access only to one-way broadband; return information is usually sent via conventional modems on telephone lines. Nevertheless, these services have been appealing enough that over one million WebTV units were sold as of October 2000. More capable STBs, such as the UltimateTV system, soon to be released by Microsoft in collaboration with DirecTV and Thomson/RCA, may have greater appeal to consumers than Web TV because they marry interactive Internet access with TiVO-like digital recording capabilities for households (Healey, 2000).

A competitor to UltimateTV will be the TV service developed by AOL Time Warner called AOLTV. AOLTV is a STB system offering an electronic programming guide (EPG), together with Internet access via America Online, for television owners. The AOLTV box is less expensive than the UltimateTV box, but by the same token offers fewer services and features.

There will be many offerings other than UltimateTV and AOLTV as the market for digital television develops in the United States. As in Europe, U.S. digital video service providers will have an incentive to create proprietary systems both to prevent nonsubscribers from accessing their systems, but also to increase switching costs from one digital service to another.

It should be added that DVD delivery of movies is another form of digital video, and the availability of DVD content together with the low cost of DVD players is priming the market for future demand for high-quality digital video content. The digital video content that is now being delivered via the Internet is constrained by the relatively small audiences that exist due to the limited deployment of broadband services to households. Bandwidth constraints and the differential speeds of PCs means that most households will only be able to view short video clips with limited pixel counts. That constraint will be greatly reduced in the not too distant future as the national DSL and cable modem networks expand.

## STREAMING MEDIA TECHNOLOGIES

The great majority of IPTV businesses use a similar set of streaming video technologies. Three firms have set market standards for streaming video: RealNetworks (Real Player), Apple (QuickTime), and Microsoft (Windows Media Player).

RealNetworks currently dominates the streaming media business. The market for streaming media services is estimated to be around $900 million. About 85% of all streaming media content on the Internet is available in formats compatible with Real Player. There are currently 155 million registered users of Real Player. The growth in the sales of RealNetworks has been over 100% per year and the company is actually earning a profit. For the year ending December 31, 1999, the company reported revenues of $131 million and a net income of $8.3 million.[1] RealNetworks has cultivated its own network of providers of Real Player compatible audio and video and has created a service called Take5 that provides quick access to the latest video content.

Over 100 million copies of Apple's QuickTime 4.0 media player have been downloaded as of October 2000.[2] Apple has created its own network of QuickTime video providers called QTV that is accessible on its Web site.[3]

Fewer copies of Microsoft's Windows Media Player have been downloaded than either Real Player or QuickTime, but most observers consider Microsoft to be a major contender in the race for future streaming media dollars and eyeballs. Microsoft has used aggressive tactics to win market share away from RealNetworks. In 1997, Microsoft purchased 10% of the equity of RealNetworks. In 1998, Rob Glaser, the CEO of RealNetworks, testified before the Senate Judiciary Committee that Microsoft's Windows Media Player had a feature that effectively disabled any version of Real Player on a user's PC without asking the user's permission. A little later, Microsoft purchased a competitor of RealNetworks named Vxtreme and then announced the sale of its equity in RealNetworks, thus producing a large drop in the latter's share price. As a result of these tactics, RealNetworks joined the coalition of companies supporting the antitrust suit filed by the Department of Justice against Microsoft.

Most major sites offer users the option to select which player they want to use at what connection speed so that they can optimize the quality of

---

[1]http://www.realnetworks.com/company/index.html?src=001101realhome_1,rnhmpg_102300,rnhmtn; http://www.realnetworks.com/company/pressroom/pr/2000/q499results.html?src=001101realhome_1,rnhmpg_102300,rnhmtn, nosrc; Amy Kover, "Is Rob Glaser for Real," *Fortune*, September 4, 2000, p. 216.

[2]http://www.apple.com/pr/library/2000/oct/10qtmomentum.html

[3]http://www.apple.com/quicktime/qtv/

video they see on their desktops. Thus, it is not necessarily the case that there will be one dominant firm in the market for streaming video services. It is likely, however, that the competition will be limited by the desire of consumers to minimize the costs connected with the coexistence of multiple market standards. Also, there still seems to be quite a bit of variance in performance of the three streaming video systems, depending on the type of content, bandwidth availability, and the performance of both the provider's and the user's systems.

## DESCRIPTIONS OF VARIOUS TYPES OF IPTV COMPANIES

IPTV providers may be divided into six categories of Web sites in order to simplify comparison of strategies: major broadcasting networks, local TV stations, large Hollywood film and TV producers, multimedia conglomerates not already covered, independent web video and animation producers, syndicators and licensers of web video. The purpose of this exercise is to look for characteristics that distinguish new and old types of content and who creates and delivers that content to final users. The line between new and old content is not always clear-cut, because the rise of cable television has already created niches for various types of nontraditional video (e.g., the *Simpsons* on the Fox Network or music videos on MTV). Similarly, the discussion looks for evidence regarding new forms of intermediation between content producers and final users that have been made possible by Internet delivery options.

### Major Broadcasting Networks

The major broadcasting networks use their Web sites primarily as a promotional or advertising device for their network offerings (see Table 14.1). ABC is experimenting with interactivity with its Enhanced TV service, aimed primarily at viewers of sports programming. These Web sites tend to be large and predictable. In the case of NBC, MSNBC (the joint venture between Microsoft and NBC) handles most of the news items, especially those requiring streaming video, whereas the NBC site seems to be headed in the direction of a general web portal. CBS, in contrast, appears to have focused on particular areas like daytime TV, news, and the top 10 lists broadcast on the *Late Show with David Letterman*.

Even though CNN and MTV are cable channels rather than national network providers, they share some of the characteristics of the major networks in this area. CNN is particularly strong in its Internet video offerings and has been something of a leader in converting its news operations from analog to digital technologies. MTV has experimented extensively with interactivity with viewers, offering comments from online chats at the bottom of the TV screen on some shows.

TABLE 14.1

*Promotional Use of the Web by Television Networks*

| Network | Web Site | Menu Items |
|---------|----------|------------|
| ABC | www.abc.com | Shows, news, and sports, Enhanced TV |
| NBC | www.nbci.com www.msnbc.com | Autos, careers, family, health, etc. (more like a web portal than the others) |
| CBS | www.cbs.com | Daytime, Late Show, news |
| FOX | www.fox.com | TV, movies, news, sports, business, kids |
| CNN | www.cnn.com | World, U.S., local, politics, weather, etc. |
| MTV | www.mtv.com | Shows, music, news, chat, etc. |

## The Web Sites of Local TV Stations

There is a relatively new and rapidly growing market for providing streaming media versions of the local news programming of local network television affiliates. Local stations also produce other kinds of content that they have made available as streaming video on the Internet. One local cable access channel, for example, made 5-minute video interviews of candidates for local political offices available on their Web site so that viewers could see them whenever they chose.

## The Web Sites of Hollywood Studios

The major studios use their Web sites to advertise new films and TV programs.

Disney appears to have separated its family content (Disney.com) from its adult content (Go.com). Disney's video content fits well with the Flash animation software owned by Macromedia, so unlike many other content producers, Disney puts all of its previews into Flash format. Most of the other studios allow the user to select a video player. The trend among studios to diversify out of TV and movie production into theme parks, tied-in merchandise, and other businesses is clear from Table 14.2. The stream-

TABLE 14.2

*Use of the Web by Movie Studios*

| Studio | Web Site | Menu Items |
|---|---|---|
| Disney | www.disney.com<br>www.go.com | Home, vacations, shopping, entertainment, etc.; Go network includes adult material |
| Time Warner and Warner Brothers | www.timewarner.com<br>www.warnerbrothers.com | Time Warner is the corporate site with links to Warner Brothers and other entertainment businesses |
| Sony and Sony Pictures Entertainment | www.sony.com<br>www.spe.sony.com | Sony is the corporate site; Sony Pictures incorporates both film and TV operations |
| MGM | www.mgm.com | Movies, television, trailers, & clips, shop, backlot |
| Fox Home Entertainment | www.foxhome.com | Store, movies, merchandise, DVD, etc. |
| DreamWorks SKG | www.dreamworks.com | Movies, video/DVD, music, TV, company |
| Paramount | www.paramount.com | Motion pictures, television, video/DVD, the studio, chat |
| Universal Pictures | www.universalstudios.com | Movies, music, theme parks, TV, home video, etc. |

ing video on these sites is of variable quality, but is designed in general to approximate in quality the trailers and clips shown on television and in movie theaters. Thus, the movie studios are likely to be early customers of services, like those offered by the Feed Room (www.thefeedrom.com) and iBeam (www.ibeam.com), that guarantee a higher level of quality of video playback on computers.

Some of the studios are experimenting with IPTV and interactivity. Disney in particular has lots of interactive web content on its site aimed at children. Paramount has a site called Entertaindom.com with web episodes of *Xena: Warrior Princess*. Sony uses its Station.com to test the market for

IPTV offerings. Still, the overall impression is that the Hollywood studios are too busy making money on feature-length narrative films to do anything truly innovative in IPTV. This may not be true in the future, especially as the potential audience for IPTV content gets into the tens of millions.

Some media conglomerates, like Hachette and Bertelsmann, do not own major film/TV studios, but are involved in a variety of related activities and are strongly involved in print media and multimedia production. Although they are not currently major producers of IPTV content, they are likely to move into this area in the future.

### Independent Web Video and Animation Producers

Here is where things get interesting. There is major growth in the number and variety of independent video and animation producers who are either trying to distribute their material through conventional channels and advertise their wares on the Internet or who create content solely for the Internet. Table 14.3 lists a few of the more interesting firms that are creating digital video for Internet delivery, but who are also selling material to other actors.

These companies are offering mostly short videos or animations aimed at an audience that finds conventional films and TV unexciting. There is often a sleaze factor to these products that appeals particularly to males in the 18- to 25-year-old cohort (e.g., the *Whip-cream Bikini Bull Riding Challenge* on Wirebreak.com and *Bikini Bandits* on atomfilms.com), but many of the offerings are high quality short films. For example, mediatrip.com had an instant success with its short satirical film, *George Lucas in Love*. The film won a number of international prizes and is currently available for purchase on Amazon.com.

Atom Films, Swankytown, and Urban Entertainment were successful in selling ideas for some of their animations to film/video distributors for TV syndication. TV and cable networks are looking for the next *Simpsons* and they seem to be relying increasingly on IPTV companies to provide forums for their talent searches. Atom Films acknowledges this explicitly in its solicitations for new material. It recruits young filmmakers from famous film schools like UCLA and USC. Atom Films recently negotiated a contract with Volkswagen of America to create 60 short videos over the next six months to appeal to younger car buyers (Volkswagen of America and AtomFilms Announce Major Content and Sponsorship Alliance, 2000).

Two companies in this category have already bit the dust: the Digital Entertainment Network (DEN) and Pop.com. The latter was the result of a partnership between DreamWorks founders Steven Spielberg and Jeffrey Katzenberg and Imagine Entertainment executives Ron Howard and Brian

TABLE 14.3

*Use of the Web by Independent Video and Animation Producers*

| Name | Web Site | Menu Items |
|------|----------|------------|
| Atom Films | www.atomfilms.com | Variety of short subjects, films sold on VHS and DVD |
| Urban Entertainment | www.urbanentertainment.com | Undercover Brother, and other animations |
| iFilm | www.ifilm.com | Great variety of short subjects; films solicited |
| Launch | www.launch.com | Mostly music videos |
| Quokka Sports | www.quokka.com | Video clips of mountain climbing and other extreme sports |
| Wirebreak.com | www.wirebreak.com | Short edgy humorous videos: e.g., Backdoor Hollywood |
| Z.com | www.z.com | Whipped cream bull-riding challenge, bobbing for maggots |
| Shockwave.com | www.shockwave.com | Animations: e.g., Joe Cartoon, South Park, Regurge, etc. |
| Swankytown.com | www.swankytown.com | Animations: e.g., Do Humans Exist? |
| MediaTrip.com | www.mediatrip.com | George Lucas in Love |

Grazer. DEN died from profligate spending on the part of its management (Lyman, 2000).

## Syndicators and Licensors of Web Video

Because of the growing demand for IPTV on business Web sites of various sorts, there is emerging a set of businesses that specialize in assembling lists of IPTV content firms and acting as intermediaries between those firms and final customers (see Table 14.4). Some of them add value be-

TABLE 14.4

*Use of the Web by Intermediaries and Production Services Firms*

| Name | Web Site | Specialization |
|---|---|---|
| ScreamingMedia | www.screamingmedia.com | Sells content of 2,800 online publishers to 1,100 Web sites |
| iSyndicate | www.isyndicate.com | Repackages IPTV content to suit customer needs in special areas: e.g., health or sports |
| YellowBrix | www.yellowbrix.com | Topic-specific news for Web sites, personalization services |
| NewsEdge | www.newsedge.com | Topic-specific news and editorial services for Web sites |
| Hitplay | www.hitplay.com | Business-to-business broadband content solutions |
| SeeItFirst | www.seeitfirst.com | Solutions to a variety of web-related problems |
| Virage | www.virage.com | Video content indexing for Web sites |
| SkyStream Networks | www.skystream.com | Network services for IPTV delivery to both PCs and TVs |

yond brokering deals by providing editorial services, licensing and copy-righting, syndication, and Web site creation/editing tools.

## CONCLUSIONS

Current IPTV content is different from current TV/cable content in being shorter, less risk averse, and potentially more entertaining to younger audiences. This is a function of the demographics of access to broadband services. Younger tech-savvy business people and students at universities with high bandwidth connectivity are clearly the target audience of most of this content.

As broadband access spreads, there may be some trend back in the direction of dominance of existing players in films, broadcasting, and cable television. Programs will be longer, although perhaps not as long as feature films. However, intermediation of talent, syndication services, production, and postproduction services will all be quite differently organized than they were only a few years ago.

There will remain important market niches for the edgy content now seen on the Internet. Because the costs of production will be much lower for small producers than they have been historically, small independent content producers will be able to survive despite the growing participation of large firms in the IPTV marketplace. As a result, there should be greater diversity of offerings in the overall video marketplace not unlike the greater diversity of audio offerings that occurred with the transition from LP records to digital compact discs.

## REFERENCES

Healey, J. (2000, June 12). Microsoft, partners introduce "ultimate TV." *San Jose Mercury News.* Source: http://www.mercurycenter.com Accessed June 14, 2000.

Lyman, R. (2000, June 9). Lights, camera, streaming video: Old-line Hollywood explores dot-coms. *New York Times.* Source: http://www.nytimes.com Accessed June 9, 2000.

Schwartz, J., with Bernoff, J., & Dorsey, M. (1999, December). *TV's Internet Tier,* p. 6. Cambridge, MA: Forrester Research.

Volkswagen of America and AtomFilms Announce Major Content and Sponsorship Alliance (2000, September 25). Source: http://www.atomfilms.com/about/press/00-25b.asp Accessed September 27, 2000.

# 15

# The Content Landscape[1]

Gali Einav
*Columbia University*

"Eventually, Television will fit on the Internet—which doesn't necessarily mean it will end up there."

—Bruce Owen

Looking at the future of interactive media, there are two main channels for delivering the next generation of television content: interactive television and Internet protocol television (IPTV). Whereas the first channel is controlled by traditional television providers and transmits content over the television screen through terrestrial, cable, or satellite technologies, the latter takes advantage of the relative freedom of Internet delivery and uses the personal computer as a main household terminal. IPTV has two classes of users: the majority who connect to the Internet using the limited bandwidth of a dial-up modem, and those who connect through cable or digital subscriber line (DSL), which is a high-speed broadband connection and provides a better quality user experience.

Definitions for IPTV vary, but they broadly refer to video content delivered over the Internet. This chapter refers to IPTV content as "Internet TV," which is streamed or downloaded and received through a personal computer. The content referred to is video, not text based, although flash animation is taken into consideration as well.

Streaming video is a technique for transferring data processed as a steady and continuous stream, allowing the client browser to start dis-

---

[1]This chapter was concluded in August 2001 and reflects the IPTV content landscape at the time.

215

playing the data before the entire file has been transmitted.[2] As such, the viewing experience is more like that of television. Once viewed, the content does not remain on the viewer's computer. Downloading refers to copying a file from an Internet server to a user's computer. This takes more time than streaming, but a downloaded file can be saved and accessed repeatedly on the PC.

Internet TV content models are based both on original programming and regenerated branded broadcast content. Borrowing from the traditional broadcast model, much of the content is video, streamed and viewed in a linear, noninteractive manner. This, however, may not be the best Internet TV content. There seems to be a need and place for the creation of new forms of content using this technology. In addressing this issue, it is useful to look at the range of content and services offered and to examine some of the main factors influencing content creation. These include the current and future Internet TV audience, technological limitations, and the economics of Internet-delivered content. Will this content justify a new revenue model in a market accustomed to accessing the web for free and, if so, how? It is hard to predict whether Internet TV will become a viable business and which content models will work, but these questions provide an interesting discussion.

## THE CURRENT LANDSCAPE OF CONTENT PROVIDERS

The majority of content providers fit into several categories (Hart, 2000); television broadcasters, meaning major networks and local and cable TV channels; large Hollywood film and TV producers; independent web video creators and syndicators, and licensers of web video. Each provides a variety of content, including news, sports, and entertainment. User-generated content is an additional popular model, in which the users provide all or part of the site's content. It is difficult to list and discuss all the content offerings available today. The following are a few examples from the current landscape.

### Television Broadcasters

All major networks offer enhancements on their Web sites that are used primarily for promotional or enhanced versions of news and sports programming. Some provide Internet-only programming. For example, ABC news (www.abcnews.com) pushes out four web-only programs accessible live or on demand, including *The Sam Show,* hosted by Sam Donaldson, "a half hour Internet-only streamed webcast featuring interviews with newsmakers and innovators" (ABC.com home page, 2001). They also offer *Internet Expose* with Chris Wallace, an ABC news exclusive program offer-

---

[2]http://thetech.pcwebopedia.com/term/s/streaming.html

ing interactive elements such as chat sessions with guests, links to related Web sites, e-mail feedback, and a world news webcast that allows chat with anchors. NBC offers two sites, www.nbci.com, a general web portal, and www.msnbc.com, a joint venture between Microsoft and NBC that provides news items, headline news, hourly updates, and videos of breaking news. CBS news (www.cbsnews.com) provides streaming versions of their top stories. On cable, CNN (www.cnn.com) offers streaming videos of headline news and top stories. Local stations, such as NY1 (www.ny1.com), offer both a streaming version of an actual broadcast as well as an archive of their news stories (including both text and video). Independent sites, such as www.bluetorch.com, work in conjunction with televised sports and provide streaming video and interactive sport programming.

## Hollywood Studios

Most studios use their Web sites to advertise new films and TV programs. Disney offers two sites. One site, disney.com (www.disney.com), is aimed at families with children, offers interactive content, including movies, TV, and video information and animation. The other site, go.com (www.go.com), has a more general audience and provides news headlines in addition to movie trailers and additional entertainment information. Sony (www.sonypictures.com) provides trailers as well as online initiatives to branded shows such as *Dawson's Desktop,* the online companion to *Dawson's Creek*. Warner Brothers (www.warnerbrothers.com) provides trailers and clips of new releases as well as classic titles. Apple (www.apple.com/trailers) offers a range of trailers from different studios.

## Independent Producers

During the past few years there have been a growing number of independent entertainment-oriented dot-coms that have been trying to find new ways to create original content, mainly short films and animation, as an alternative to conventional television programming. Some of these sites, such as television.com and breakTV.com, provide access to previously broadcast television shows as well as behind the scenes interviews and streaming video highlights of brand name network and syndicated shows. Others provide original programming and access to numerous short films. One of the best known examples for this genre is pseudo.com, which featured original short films and programming, but was forced to shut down in the fall of 2000. IFILM (www.ifilm.com) provides an online video on demand portal with over 80,000 films, including news reviews, trailers, and a video shopping guide. It also produces a weekly television series in partnership with "The Independent Channel," covering the world of independent film. AtomShockwave, a merger between Atomfilms and

Shockwave (www.atomfilm.com), provides a huge pool of game, film, and animation content online for consumers and businesses.

MediaTrip (www.mediatrip.com) is an entertainment portal providing on-demand film, music, and original programming content for adults from ages 18 to 49. Hypnotic.com, which merged with nibblebox.com, creates vector animation (Flash) deliverable over low bandwidth; distantcorners.com provides science fiction shorts and has recently formed a partnership with Sony pictures digital entertainment. Icebox.com, heavy.com, mondo media, and wildbrain.com primarily create and distribute animation. Some of the best-known creations of this genre are *The God & Devil Show* (mondo media) and *Mr. Wong* (Icebox). Wirebreak.com develops and produces programming for distribution on television video or other Web sites. The site voxxy.com provides shows for teenage girls, including a series with Jennifer Aniston.

## Content Syndicators and Licensors

Some sites specialize in assembling lists of content firms and act as intermediaries between those firms and final customers (Hart, 2000). Examples include Screamingmedia (www.screamingmedia.com), which sells the content of 2,800 online publishers to 1,100 Web sites; Virage (www.virage.com), which is video content indexing for Web sites that provide video owners with Internet content distribution solutions; and iSyndicate (www.isyndicate.com), which enables the collection, distribution, and management of content across the Internet. The Feedroom (www.feedroom.com) aggregates predominately news content from nationwide sites, including 13 NBC channels, 17 Tribune channels, Reuters, four Journal stations, and one Granite Group station.

## User-Generated Content

In this model, the content providers are the users themselves. The viewers create and send their own videos for transmission over the web, creating personalized channels that can be shared with users worldwide. This model appeals to the public, which can broadcast its own content and thereby affect the programming. It also enhances the business proposition by providing a cost-effective way to produce, acquire, and market content (Morisano, 2001). Examples include alltrue.com, an entertainment platform in which the users can watch, collect, and send video clips that are usually reality based. Sportscapsule.com allows teams to upload their videos of local sporting events, which the company then enhances with popular music, graphics, and prerecorded voiceover comments from sport personalities such as ESPN's Chris Berman and TV football commentator John Madden. Earthcam.com (http://tv.earthcam.com) allows users to create their own "personal TV channel" on the PC and to broadcast content to friends, family, or the World Wide Web. Anivision.com enables a three-dimensional interactive viewing experience that allows viewers to direct their own productions.

## PROMINENT REASONS CONTRIBUTED TO THE FALL

"Everyone had great ideas—it was just before their time."
—Peter Scott, Nascar.com

Only a short time ago, the excitement over the vast opportunities offered by the web reached a peak. In December 1999, a Forrester report predicted that broadband penetration would rise from six million broadband users in 2000 to 19 million in 2002 (Schwartz, 1999), more and more of them young Internet audiences.[3] Other advancements, such as the new capabilities that streaming technologies offered for innovative programming, helped create an air of excitement around Internet TV. The past year has seen the fall of many Internet content ventures. Expressions such as "Black September," referring to September 2000, saw many of these companies—including the Digital Entertainment Network (den.com), pop.com, and pseudo. com—forced to close their doors (Hollywood Flops, 2000). When David Wertheimer, chairman of Wirebreak.com, coined the phrase "Hollywood's Vietnam," he was referring to the phenomenon of people rushing into Internet businesses without a clear idea of why they are getting in. This scenario depicts the grave atmosphere surrounding this industry (Lyman, 2000).

### Business Models Undercut by Disappointing Broadband Penetration

By summer 2001, only 2 out of 10 million DSL phone lines that had been predicted for the United States were actually installed. Although 70 million of the country's 105 million households have access to cable TV, and about 60 million have access to cable modems, less than 10% of those have signed up for the service (Yankee Group Report: Broadband—What Happened?, 2000). Most of the entertainment sites catered to a broadband audience, but there were too few broadband users to sustain viable businesses. Most of these companies did not have a business model that included additional revenue streams. The majority of the web savvy audience in the United States, which was accustomed to receiving content for free, would not pay for the new Internet content. At the same time, due to the dot-com crash, both investors and advertisers pulled out of the Internet, resulting in substantial financial losses and making it impossible

---

[3]According to the same Forrester report, young consumers are 29% more interested in broadband than their adult counterparts. See Schwartz, 1999. *TV's Internet Tier*, p. 4.

for many of these companies to survive. The Digital Entertainment Network (DEN), for example, which created 4- to 6-minute streaming videos as well as 13 online series ranging from sports and music to sex-filled drama, were targeting such a broadband audience. After raising $50 million from companies, including Microsoft and PepsiCo, and paying its top eight executives excessive salaries described as "Hollywood excess meets Internet euphoria" (Digital Entertainment Network: Start up or Non-Starter, 1999), the company lost $27.1 million in 18 months and closed in October 2000. Pop.Com, a joint venture formed by DreamWorks' Steven Spielberg and Imagine Entertainment's Paul Allen, laid off most of its 70 staffers in early September 2000 and indefinitely postponed the launch of the pop.com site. According to DreamWorks' Jeffrey Katzenberg, even though it had a very strong financial backing, the company understood that there were not enough people willing to pay for their content and that the high operating costs, estimated at $2.25 million a month, "would probably be thrown away" (HollyWeb Flops, 2000).

## Technological Issues

Video content takes a long time to download. In January 2000, Miramax's "Gunivere" was the first Hollywood movie to be offered online as a download in a legal, nonpirated way. The download time, through a DSL connection, was 1 hour and 14 minutes (Tristam, 2001). A new initiative of five Hollywood studios (MGM, Paramount, Sony, Universal, and Warner Brothers) provides on-demand access to their films and promises a shorter download time of 20 to 40 minutes per film (Umstead, 2001). In addition, the quality of the stream available to an individual user can be negatively affected by other users' requesting the same content simultaneously. This usually leads to poor quality video, as well as net congestion and delays that turn the viewing experience into a disappointing one. Users accustomed to viewing video on a television or movie screen, neither of which crash, stall, or take long minutes to download, find this hard to accept. Companies seek to overcome these problems by providing technology that ensures that only the highest quality streams reach their audiences. Akamai's "Steadystream" (www. akamai.com), for example, handles the quality of live broadcasts by sending multiple copies of the video to the edges of the network, closer to the viewer, to ensure that a good quality video stream reaches the viewer.

## Poor Fit Between Content and Audience

One of the main questions for Internet content creation is audience behavior. Many content providers failed to determine in advance whether their target audience was interested in viewing entertainment content on a PC instead of on a television or a movie screen. Perhaps other forms of content, such as news and business information, would be more suitable for the web. In 2000, 72% of broadband users had this access at their workplace, 8% were at a school or in a library, and only 20% had broadband at home (Carey, 2001b). It seems that entertainment might not have been the optimal form of content for an audience at work that cannot devote long periods of time to viewing video content. Much of the early video content on the web, including *Froggy in the Blender* and *Bikini Bandits,*[4] which appealed to the young 18- to 25-year-old male audience, did not cater to the mass business audience or the growing numbers (from 14% in 1994 to 51.4% in 2001) of female viewers (Carey, 2001a). As a comparison, financial news sites like Bloomberg.com, which target the workforce with financial information, as well as MSNBC.com and CNN.com that provide news stories for both the home and working audience, are reporting steadily growing traffic.

## Costs Per Production

Streaming video is available today through several vendors, including RealPlayer, Apple Quick Time, and Microsoft Windows Media Player. All three players are offered as free downloads or bundled with consumer software. The situation is different for the content provider. Streaming Internet television content is currently more expensive than delivering similar content by cable or satellite (Waterman, 2000). In addition to fixed capital costs for encoding and storing video, there is also a high variable cost. In broadcast, costs are on a per program basis, so the more viewers you get the more money you make. Internet transmission is the opposite. The more people who view a program, the more expensive it gets for the content provider. The cost for the content provider can be divided into three areas: hosting the content, streaming the content, and the cost of broadband. Hosting and encoding are a fixed cost. The more content hosted and encoded, the larger its file size and the more the content provider will pay the Internet service provider (ISP). A 3-hour movie costs more to host and encode than a 1-hour movie. But once a certain size is determined, these costs are fixed and remain the same whether 10 or 100 people view the program. The variable costs are for streaming and bandwidth. Although revenue deals vary, the rate is determined per stream, so the more streams requested the more expensive it gets. The cost of the bandwidth will grow according to the number of users as well. This creates an ironic situation because the more successful in terms of the number of viewers the Web site is, the less profit it makes. For example, live broadcasts that are popular with Internet audiences are usually not commercially viable. While advertisers pay to be part of a successful Web site, the fixed cost of the advertisement does not grow according to the number of users. In contrast, the provider's costs continue

---

[4]Available through ATOM Films. www.atomfilms.com

to increase as their audience grows. Although advertisers do not pay per user, streaming content providers do. This does not mean live broadcasts do not exist. Microsoft Network (MSN) picked up the tab for Madonna's half-hour web show, which was streamed in November 1999, and generated nine million streams. MSN rationalized that the publicity generated was worth the expense and was balanced by their not being required to pay for the concert rights (Lassiter, 2001).

## SURVIVING THE FALL

"I am an optimist and I want to believe some of these companies will break out. The odds are slim, but I'm still hopeful, which is what makes it fun"
—Kenneth Wong, CEO of the former pop.com

Despite the bumps and hurdles along the way, the Internet TV landscape continues to develop. Companies have learned from previous mistakes and are developing new models to get them through this period. Companies that have strong financial backing and branding, including the networks and established Hollywood production companies (e.g., Bloomberg, CNN, and Sony), could afford to absorb the loss from their web divisions and were less affected by the Internet downturn. There are several characteristic approaches taken by companies that managed to stay in the game. Some chose to focus on a single approach and others combined several as a survival strategy. Some companies were able to diversify the high cost of producing content solely for the Internet by creating content that could be used on several platforms, and thereby were able to generate additional revenues. The other platforms include television, video, PDAs, and wireless devices. Other companies reverted to providing content through narrowband dial-up connections.[5] Most news sites offer both narrow and broadband options for accessing video. Still others, understanding the limitations and advantages of the Internet, revised their business models, including their programming, promotion, and transmission solutions to better fit the audience and cut bandwidth and streaming costs. This allowed them to rely less heavily on advertisers and investors. These companies turned to licensing and distribution deals, subscription or premium pay services, and innovative advertising models. Of course, finding the content that people will pay for remains one of the main challenges facing the industry.

### Companies Adapted to Changes in Their Environment

Nascar.com decided to branch out from a sports-only site to an entertainment site in order to create a "buzz" and attract a new entertainment audi-

---

[5]According to a Jupiter Report, two thirds of Internet users will still be using dial-up connections in 2005. See Tristam, 2001.

ence. In the near future, the site will offer a weekly 3-minute-long cartoon, by the creators of *The Simpsons*, that will pay homage to Nascar. As a revenue generating advertising model, product placement will be used throughout the cartoon. Nascar will also explore a new model for unique premium content in which the customer experiences the illusion of participating in a real race by following, on their PC screen, a video camera installed in the race car itself. Promotion will be done over both the TV and the Internet, with heavy promotion from AOL.[6]

Interactive television (INTV) (www.intv.tv) and icebox.com provide examples of two entertainment Internet startups that were forced to close and have now revised their business models in preparation for relaunch. INTV bought former pseudo.com, including its content, for $1.8 million in January 2001. The investment was recovered by selling hundreds of computers and renting out the former Pseudo facilities (including its television studio). In order to cut down costs, staff was drastically cut from 200 to 5. College interns and volunteers are filling vacant positions. In order not to rely on outside investors and advertisers to keep the company running, INTV is now adopting a new subscription revenue model. Along with content that will be offered for free, access to the popular Pseudo content will be available for a monthly subscription of approximately $5. New shows cater to a 15- to 30-year-old New York–based audience and include a version of *American Bandstand* streamed live from the Wetlands club in New York City. They are seeking to recapture their female audience by bringing back Pseudo's *Cherry Bomb,* a show produced for women by women. Programming will take into consideration the "at work" audience, which peaks during lunch hour. In order to cut down the broadband and streaming costs, INTV is working with partners to close deals with streaming vendors that have bandwidth surpluses. Instead of spending money in advance, they plan to grow slowly as the industry evolves.[7]

Icebox.com was founded by successful television writers who felt frustrated in the TV process with their lack of control over the final product. They thought the Internet would be a great platform to create cheaper animated shows that could eventually be migrated to the television. Series such as *Mr. Wong,* which garnered three million viewings of a particular episode, proved very successful. They were able to break even by using flash animation and other techniques to keep their file sizes small, providing a higher quality user experience on low speed connections. Despite their relative success, Icebox was forced to shut down in February 2001 because advertisers and investors pulled out of the Internet. Like INTV, they reduced their

---

[6]Personal communication with Peter Scott, Senior Director of Multi Media Content for Nascar.com, Interview, August 9, 2001.

[7]Personal communication with Edward Salzano, CEO/CTO INTV Inc., Interview, August 22, 2001.

staff, sold assets to pay debts, and decided not to rely on advertisers. They relaunched the company on May 16, 2001. Icebox no longer produces content without a sponsor or production partner. In addition, they are initiating an on-demand model requesting users to pay from 25 to 50 cents per show. They believe this VOD model will keep the site running and have posted an official explanation for these changes on their Web site. Icebox intends to focus on the development and exploitation of their content and believes that revenue will come from production and development deals as well as the online and offline syndication of their web content, such as their syndication deal with Mondo Media. Because the animation produced is in small files, it can be licensed and delivered on other platforms including wireless devices, video, DVD, and broadcast television.[8]

Heavy.com offers a mix of alternative music and humorous video clips as both free and premium content. The premium content is available for $7 a month or a $50 annual fee. The site is not necessarily intended to be profitable; it is viewed as a way to broadcast television over the Internet without a broadcast license and a distribution network. The intent is that this content would later be broadcast on television to generate more revenue.[9]

Romp.com, which targets male audiences, gave the subscription model a try at $34.95 a year, but decided it was not sufficiently successful. They plan to refund all their subscribers and change their business model. Romp.com now plans to develop branded film and print objects, stop updating content daily, and release new shows periodically.[10]

AtomFilms is in the short film distribution business and not necessarily the Internet streaming business. In fact, 60% of their revenue comes from selling shorts to airlines and shopping malls. They also signed a $1 million distribution deal with Blockbuster.com. Atom Films has deals to air short films in hotel rooms and wireless devices from companies, including Compaq, Sanyo, and Texas Instruments (HollyWeb Flops, 2000). In January 2001, AtomFilms and Shockwave.com, which features interactive games, merged. The two companies intended to build vast online entertainment content that, due to the dot-com decline, was postponed. They also cut their staff from 180 to 30 employees and in May 2001 announced a pay to play initiative featuring two online games packages at $19.95 and $29.95 (Olsen, 2001). MediaTrip signed a distribution arrangement with Amazon.com and sold their successful short film *George Lucas in Love* (a parody of *Shakespeare in Love*) through the retail outlets Tower and Blockbuster. By January 2001, Amazon had sold 20,000 tapes for $7.99 (DVDs are available for $12.99) and 25,000 more have been bought in stores (Hart, 2001).

---

[8]Personal communication with Tal Vigdersom, Managing Director of Icebox.com, Interview, August 17, 2001.

[9]Assad Simon, cofounder of Heavy.com (see Tristam, 2001).

[10]Information available on Web site, www.romp.com

Many companies are working on deals to sell Hollywood studios the rights to their content. One of the biggest deals is the $2 million that Universal Studios Inc. paid UrbanEntertainment.com (www.urbanentertainment.com), a site that caters to the African American market for the movie rights to the animated short *Undercover Brother* (HollyWeb Flops, 2000).

Even companies with strong financial backing are developing new business models. MSNBC.com started to stream video as early as 1998. To cut production costs, materials left on the cutting room floor at NBC and MSNBC cable were used for the site. Because half of the MSNBC.com audience is at work and the other half logs on from home, the site offers both broadband and dial-up options. To keep costs down, MSNBC does not stream high resolution video. In addition, they are planning to branch out to PDAs and wireless devices. MSNBC.com will be launching a video player that will allow advertising links from banners, as well as a play list that enables the viewers to select additional videos.[11]

## SELECTING SUITABLE CONTENT

"Why do things on the Web if it's just like watching a TV show?"
—David Wertheimer, President of Wirebreak.com

Providing suitable content given the limitations of the web is not an easy task. Content needs to be innovative, compelling, and suitable for a PC screen as well as have the ability to create a connection that will bring users back to the site. The majority of the audience is at the workplace, so short content, under 15-minute segments might be suitable.[12] Sites that provide news content to mass audiences that are not necessarily early technology adopters might provide for both high and low speed connectivity.

Animation may work better than video because it can be less expensive to produce and, as the files are smaller, it can be streamed more efficiently over both narrow and broadband lines. Animation works well on many platforms, so it is easy to localize and redistribute. Interactivity and nonlinearity work best for Internet TV content. It is therefore likely to appeal to a "lean in" crowd, characterized as an active one-on-one personal experience associated with a computer in contrast to a "lean back" passive audience associated with television viewing. Broadband is very important for this content form. Low quality video and slow connections hamper the quality of most sites that stream video.

---

[11]Personal communication with Michael Silberman, Executive Director MSNBC, Interview, August 24, 2001.
[12]34"405", a short, 3-minute film created on home computers by Bruce Branit and Jeremy Hunt, is a great example. Available on ifilms.com

## Entertainment

Most of the innovative Internet TV content will continue to come from the entertainment sector. One interesting entertainment site that takes advantage of the Internet's capabilities is Sony Screenblast (www.screenblast. com), an extensive broadband entertainment portal for films and music buffs. With this site, Sony seeks to strengthen its two-way relationship with the audience and to get them more involved. Sony's goal and revenue model is to provide the audience with sufficiently compelling content to stimulate software sales and subscriptions. They also intend to attract integrated advertising through personalization and customization. The site caters to the 18- to 24-year-old audience, providing tutorials and instructions on how to create feature films, including special effects, that can be uploaded to their friends or a Sony producer for feedback. Whereas the free membership allows the user 50 Mb of storage space, six editing tools and free trials, a full deluxe set of these of tools is offered for $169 and can be purchased separately. The strategy is to build long-term relationships with customers that will come back for additional tools, instructions, and feedback. As for developing content on other platforms, Sony is waiting to learn what kind of entertainment experience belongs on each platform.[13]

Internet TV entertainment may become intertwined with advertising and commerce becoming advertising-sponsored content. BMWfilms (www.bmwfilms.com) provides an interesting example of using entertainment for advertising. BMW research shows that 85% of their consumers go online before making a purchase and because they were interested in creating a new and different branding campaign, BMW decided to use the Internet. In April 2001, they launched a site that features five short films, less than 7 minutes each, directed by intriguing names such as Guy Ritchie, featuring his wife Madonna, and Ang Lee. All films star a BMW car and are shot at cinematic quality using 35 mm film and high production standards. The concept was pitched to the directors as an opportunity to make quality entertainment for the Internet with complete artistic freedom. The films are offered in both broadband and dial-up versions and are promoted through traditional broadcast commercials and Internet shorts sites. BMWfilms has been approached by TiVo to offer its viewers broadcast quality versions of the films on their television sets. There is no known direct link between the launching of the site and an increase in BMW sales. But the fact that users who log on, disclose information, and may choose to be contacted combined with the great buzz BMW is receiving, looks like successful advertising.[14]

---

[13]Personal communication with Andrew Schneider, Senior Vice President for Broadband Content, Sony Pictures Digital Entertainment, Interview, August 24, 2001.

[14]Personal communication with Karen Vonder-Meulen, Marketing & Events Communications Manager, BMW, Interview, August 22, 2001.

## News and Sports

News stories are relatively short and their audience seems to be more forgiving when it comes to the quality of news-oriented video content streamed over a dial-up connection. News is a natural for interactivity and has a market of interested viewers who gain the prerogative of receiving news and updates at their convenience. News sites are reporting a steady increase in audience. According to Bloomberg statistics, for example, users are staying an average of 36 minutes on the site.[15] Breaking news, is the most popular content and the biggest in terms of video streams for MSNBC.com. The day of the Seattle earthquake, the site sent 1.5 million streams on demand. This is close to a cable size audience. Hourly updates of video news clips and web exclusive video headlines are also very popular because news can change hour by hour. On average, MSNBC handles 200,000 streams per day and offers content in both broadband and narrowband versions.[16] The Feedroom reports 2.5 million streams a month. They found that streamed local news on the site added a new predominantly male audience that logged on from work.[17] These viewers, added to their female homemaker-based audience, made the site appealing to advertisers.[18]

Some Internet TV models provide interactivity and a community around sports. Sports fans are a loyal audience more likely to pay for additional content. NBA Entertainment (www.NBA.com), for example, has become one of the most popular sports sites, averaging more than 800,000 daily visitors throughout the NBA finals 2000. The site offers access to all 29 of the team's Web sites and an NBA store. In addition, it has interactive features such as custom headlines of favorite teams or the option to create individualized packages of video and audio clips (Kaufman, 2001).

## Children's Programming

Children are becoming a natural audience for interactivity because they are accustomed to it from a very young age. HBO's *The Deadwood Mys-*

---

[15]Personal communication with Michelle O'Brien, Head of ITV&BB Division, Bloomberg.Com, Interview, August 20, 2001.

[16]Personal communication with Michael Silberman, Executive Director MSNBC, Interview, August 20, 2001.

[17]The numbers provided in this chapter changed after September 11, 2001. In a later interview, Silberman added that during September 11, MSNBC had record traffic of approximately 12 million unique visitors to the MSNBC site. The video traffic that day was also a record—approximately 6.3 million live streams and another 5.75 million on-demand video streams were requested.

[18]Personal communication with G. Gooder, Manager Business Development/The FeedRoom, Interview, August 14, 2001.

*teries* (www.hbofamily.com/deadwood) is a 16-episode series that debuted on July 16, 2001, and ran through Halloween. It was one of the first original children's programs created for the web. Situated in Deadwood, Oregon, the series focused on the search for Jessica Fischer, who mysteriously disappeared on her 16th birthday. Jessica's sister Rachel and three friends tried to solve the mystery with the help of the home users. Each Monday a new episode with videos of the characters, clues, evidence, reports, and links to FBI files aired, and children were able to send in their thoughts via e-mail. Every few weeks, an online chat brought together the avid followers of the series. The program aired, despite the crash of the dot-com market, because funds for it had already been committed, and the HBO producers are happy with the results. It took the team 4 months to transform the concept, created by television producer Andre Mika, from the passivity of television broadcasting to the interactivity of the Internet. This included cutting down the original proposal from 13 clips per episode to 3 or 4, and to making full use of multimedia. The intended age group was 10- to 14-year-olds. Even though it is hard tracking user information on children, HBO research shows that, on average, kids stay online for five hours and keep coming back. The site has already created a sense of community and a fan club has been created in Staten Island. It seems that children crave this kind of interactivity.[19]

### Information-Based Shows

Information-based programs, such as documentaries, work well with web enhancements because the Internet gives viewers the option to pause a show and look for complementary information. Public television, a natural home for this form of broadcasting, is providing pioneering content. WGBH, the Boston-based PBS station, is working on an interactive documentary named *The Commanding Heights* to be aired in April 2001. The show producers, Howard Cutler, Frontline producer Mike Sullivan, and Pulitzer prize winner Dan Yurgen have decided to create a parallel production from ground level, which is a collaboration between television and web producers. This means shooting on site will be planned in advance for both the TV program and web enhancements, a production method that is both efficient and cost effective. The entire show will be streamed via broadband. There will be a window of about a month to stream a show to allow PBS to sell the video after it airs.[20]

---

[19]Personal communication with Lynne Eyberg, Co-executive in Charge of Production; and Noa Morag, Web Producer, HBO Family, Interview, August 9, 2001.

[20]Personal communication with Curtis Wong, Manager, Microsoft Research, Interview, August 20, 2001.

## Education and Training

Many educational organizations—including universities, forums, and training programs—offer content on the Internet. More and more sites provide a video component of the lesson itself. Distance learning is becoming a popular option for accessing course information unconstrained by geography and schedules. Education seems to be a natural use of the web, because the interactivity is ideal for guiding users, teachers, and students, through content as well as creating learning communities. The site "e-school online," for example, created by ACTV, a New York–based creator of proprietary and patented software tools, instructs grade school teachers on how to teach reading. Because not all schools have the bandwidth needed for streaming, the same content is distributed both by video stream and by CD-ROM.[21] The NASA Education program (www.education.nasa.gov) is a gateway to a wide variety of NASA Web sites for teachers and students. This content model can work in both an educational and business environment. It is expensive for people to leave their offices to attend classes, so training on the web is often a good solution for companies and corporations. It allows them to both cut costs and enhance the work environment.

## Corporate Communications

It is common for corporations to have a video communication department that creates content for the internal use of the company, including business meetings, and may provide educational and instructional videos. Some independent dot-coms, such as the FeedRoom, provide companies with an option to outsource content of this sort. The Feedroom receives tapes from a company and adds metadata links and text to create a streaming asset available to the company's employees, the media, or investors.[22] Companies like Cisco provide services that split streams, allowing them to reach more people. This allows companies to cost effectively send one broadcast stream to thousands.[23]

Many believe that the future of Internet TV content lies in the business sector and the ability to webcast announcements over private corporate intranets, accessible to employees, investors, customers, and the press. According to Blake Hayunga, CEO of Street fusion, webcasts have more than doubled in the past year. Webcasting is cheaper than a dial-in confer-

---

[21]Personal communication with Craig Ullman, Chief Creative Officer, ACTV, Interview, August 26, 2001.

[22]Personal communication with G. Gooder, Manager Business Development, The FeedRoom, Interview, August 14, 2001.

[23]Personal communication with Frank Scibilia, Product Manager, Cisco Systems, Interview, August 23, 2001.

ence call. Pricing ranges from \$1,000 for an audio stream to \$8,000 for a video stream. Ten annual webcasts cost roughly the same as one dial-in conference call. At approximately 30 cents a person per minute, conference calls in which large numbers of people participate become expensive (Arora, 2001).

## Pornography and Games

Pornography and games are two very substantial topics, too large to discuss in confines of this chapter. In passing, however, it should be mentioned that pornography is the largest revenue generating content on the web today and even "soft" sites such as naked news.com are offered on a paying basis.

Providing video on the Internet for games played on television may contribute to an increase in broadband penetration and Internet TV content,[24] but because this is only speculation it also falls outside the scope of this discussion.

## THE FUTURE OF INTERNET TV

Changing people's viewing habits is hard to do. It takes time to understand the characteristics of a new medium and create exciting content for it (Carey, 2001b). Views about the future of Internet TV range from pessimists, arguing both that broadband is not yet here and the web is not suitable for broadcasting video, to optimists that argue that Internet infrastructure is the most suitable for VOD and interactive content. The Internet is growing by at least 85 million users per year. Webcast content reaches more than 50% of these users and should reach about 475 million users by the end of 2001 (MRG Multimedia Research, 2001, and Arbitron and Edison Media Research, 2001). Harris Interactive Research found that the number of U.S. households with broadband connections grew by 41% between April 2000 and January 2001, and cable modem services only increased about 10% in that same time period (Stanfield, 2001). There are 350 million secure Windows media players that have been distributed, and 215 million Realplayer users worldwide.[25] Reports show that video streaming on the net grew 215% in 2000 to over 900 million streams (DFC, 2001a).

## Will Internet TV Be Profitable?

There is a general consensus that unless some fundamental shift occurs and people begin to pay for the content, Internet TV will never be profitable. As the

---

[24]Personal communication with Rob Davis, Executive Producer *SpiderDance,* Interview, August 16, 2001.

[25]Based on information given by both company's sales and marketing department.

number of broadband users and streams continue to grow, with an average of 1.2 video advertising opportunities per stream (DFC, 2001b), advertisers are beginning to take notice. This may lead to a change from the old advertising model of buttons and banners to a new targeted advertising model that can provide a solution for advertisers that are losing their broadcast audience. Inserting "in show" commercials before webisodes, instant online purchases of products, and the promotion of companies through entertainment, such as BMW Films, may prove suitable for Internet TV. The return to the 1950s model sponsored shows and segments may also work. Using Hollywood content available for syndication and international distribution may prove an important source of revenue as well (Waterman, 2000). Small niche markets, such as travelers, independent filmmakers, and other special interest groups might pay for content that is not otherwise available.

Although it is hard to pinpoint the number of users needed to justify the investment, some number of people may be willing to pay small amounts for compelling content. There are already examples of paying models that work. The WWF (World Wide Film) site directs viewers from their cable pay per view events to the Internet, where viewers can continue to watch and pay. House of Blues (www.hob.com) charges $4.99 for high quality live festivals and iLive.com offers pay-per-view shows for $1.00 to $3.00 per show. In order to create an incentive for users to tell family and friends about the shows and to build a wider base of paying users, iLive.com offers users 25% of the revenue made from the shows their friends decide to watch. Real Goldpass offers a $9.95 monthly membership that allows access to premium content such as 24/7 live coverage of CBS's *Big Brother* house, ABC's Connie Chung's interview with Gary Condit, and adult content such as *Bikini Fever*. Real Goldpass reports more than 300,000 paying subscribers. This might not be a large number, but it may mean that cracks are beginning to appear in the general perception that all content on the Internet should be free. Still there are those in the industry that believe that until the technology provides better quality streaming, many customers will refuse to pay. Only time will tell which direction the pay model might take. If the cost of broadband and streaming goes down, the current billing model of pay per stream changes, and the number of broadband users goes up, then Internet TV content providers may succeed.

According to Professor V. Michael Bove[26], "People tend to confuse a delivery mechanism with an audience and in the long run there will be no distinction." In the future, the distinction between Internet TV and interactive television may blur. As a result, content and revenue models may do so as well. Whereas some of the less popular channels on cable might move to the Internet, streamed content can find its way to the television screen with no recognizable visual difference.

---

[26]Personal communication with Professor V. Michael Bove, Object Based Media Lab, MIT, July 7, 2001.

The Radon video card by ATI technologies and Sony's Vail Digital studio are two new products in the works that will enable viewers to watch TV via their PCs. Companies like INTV are building on the ability to view IPTV, TV, cable, and video on a PC, which is their preferred future content and delivery platform. In general, most content providers are interested in migrating their content to a one screen experience. This means viewing and interacting with content on a television screen. Interesting steps are being taken in this direction with the deployment of the Cablevision set-top box and the joint Comcast and Scientific Atlanta VOD service. Currently, the technology and deployment of set-top boxes tends to be more costly and less robust than the Internet, but Internet TV programs are providing a learning ground for future one screen interactivity.

## CONCLUSIONS

The question of how Internet TV will fit into the overall television landscape in the future is both intriguing and of importance to the media industry. Lessons learned include the importance of broadband, recognizing and targeting potential audiences, and the necessity of suitable content and business models. Innovative Internet TV content that employs creativity and versatility will benefit from increased broadband penetration. Entertainment dot-coms have cut staff and production costs and are waiting for the technology to catch up. In order to ease the need to rely on advertisers and investors, new revenue models for on-demand content, premium services, and subscriptions are being implemented. Advertising, meanwhile, is moving toward a branding and sponsorship model.

The number of broadband users is growing slowly but steadily and potential audiences are showing interest in both existing and new streaming content. Internet TV may prove an alternative mode of broadcasting for independent content creators that have too small an audience for broadcast. Independents may capture the lost TV audience that has turned to the web.

Internet TV differs from traditional TV, so content and business models should be adjusted accordingly. Two main challenges for Internet TV are to change traditional television viewing habits and to convince viewers to pay for Internet TV content as they do now for cable and satellite programs. Understanding the intended audience and creating compelling, interactive, exclusive content can drive these changes. Creating content deployable on multiple platforms can diversify production costs and increase revenue channels until the costs for broadband and streaming come down. It is anticipated that interactive programming will migrate to a one-screen television experience, combining the interactive capabilities of the Internet with the viewing experience of television.

## ACKNOWLEDGMENTS

Professor V. Michael Bove, MIT Media Lab; Vladimir Edeleman and Fank Barbieri, Filter Media; Lynne Eyberg and Noa Morag, HBO Family; G. Gooder, The FeedRoom; Rob David, Spiderdance; Dan Dubno, CBSnews.com; Michelle O'Brien, Bloomberg.com; Edward Salzano, INTV, Inc.; Andrew Schneider, Sony Pictures Digital Entertainment; Frank Scibilia, Cisco Systems; Peter Scott, Nascar.com; Michael Silberman, MSNBC; Craig Ullman, ACTV; Tal Vigdersom, Icebox.com; Karen Vonder-Meulen, BMW Films; Curtis Wong, Microsoft Research; Marcia Zellers, AFI.

## REFERENCES

ABC.com homepage (2001). Retrieved August 1, 2001 from www.abc.com

Arora, A. (2001, May 7). Talk is cheap. *The Industry Standard.*

Carey, J. (2001a). *Demand for new media.* Lecture presented at Columbia Business School, June 6, 2001. New York.

Carey, J. (2001b). Audience demand for TV over the internet, p. 7. Retrieved August 1, 2001 from www.gsb.columbia.edu/faculty/jcarey

DFC Intelligence Report (2001a, January). DFC press release. Retrieved September 1, 2001 from www.dfcint.com/news/prJAN092001.html

DFC Intelligence Report (2001b, June). DFC press release. Retrieved September 1, 2001 from www.dfcint.com/news/prJUNE272001.html

Digital entertainment network: Start up or non-starter? (1999, November 15). *Business Week Online.* Retrieved August 1, 2001 from www.businessweek.com

Hart, J. (2000, November). Content Models: Will IPTV be the same or different? Retrieved August 1, 2001 from www.citi.columbia.edu/abstracts.htm

Hart, M. (2001, January 14). A comeback for short films is linked to the Web. *The New York Times* [online version]. Retrieved August 1, 2001 from www.NYTimes.com

HollyWeb flops. (2000, October 23). *Business Week Online.* Retrieved August 1, 2000, from www.businessweek.com

Internet VII: The internet and streaming: What consumers want next (2001, August 23, 2001). *MRG Multimedia Research Group, IP Video and Streaming Media 2001 Market Forecast 2001–2004, MRG, INC., and Arbitrion and Edison Media Research,* Adweek IQ morning briefing. Retrieved August 23, 2001 from www.adweek.com

Kaufman, D. (2001, April). The Web's got games. *Digital TV.*

Lassiter, T. (2001, February). Internet not ready for prime time. *Digital TV,* p. 50–51.

Lyman, R. (2000, September 6). Light, camera, streaming video: Traditional Hollywood exploring dot-com entertainment. *The New York*

*Times* [online version]. Retrieved August 1, 2001 from www.NY Times.com

Morisano, J. (2001). Alltrue Networks press release. Retrieved August 1, 2001 from www.alltrue.com

Olsen, S. (2001, August 21). AtomShockwave puts on a new face. *CNETNews.com*. Retrieved August 21, 2001 from www.CNETnews.com

Schwartz, J., with Bernoff, J., & Dorsey, M. (1999, December). *TV's internet tier* (Forrester Report). Retrieved August 1, 2001 from www.forrester.com

Stanfield, S. (2001, March). Streaming for cash. *Digital TV.*

Tristram, C. (2001, June). Broadband's coming attractions. *Technology Review*. Retrieved August 1, 2001 from www.technology review.com/magazine/June 01/tristram.asp

Umstead, T. (2001, August 20). Studio team up for VOD via the web. *Multichannel News.*

Waterman, D. (2000, November 29). *Economic models for internet TV content providers*. Paper prepared for TV Over the Internet conference. Retrieved August 1, 2001 from www.citi.columbia.edu/abstracts.htm

Yankee group report: Broadband—what happened? (2000, June 11). *Business Week Online*. Retrieved August 1, 2001 from www.businessweek.com

# Future Impacts

# 16

# Will Internet TV Be American?

Eli Noam
*Columbia Institute for Tele-Information*

For several centuries, culture flowed largely in one direction: out of Europe to the colonies and the rest of the world. Then, after World War I, the flow reversed direction for the young medium of film. Around the world, audiences flocked to Hollywood movies. European cultural elites, shocked at the loss of control over their publics, led a countercharge. They promoted protectionism to support centuries-old national cultures against a few vaudeville theater promoters who had pitched their tents in Hollywood. But despite seven decades of efforts, this challenge remained. In 1999, of the 50 highest grossing films worldwide, 49 were American. The year before, it was 22 of the top 25. In Germany, domestic films were down to 10% of the audience (the rest was predominantly for American films). In the United Kingdom, the domestic share fell to 14% in 1998. Even in France, the audience share for domestic productions has dropped below one third of the total. The European Union, to stem the tide, provided subsidies of $850 million for film, but they generated box office revenues of only $400 millions.

After World War II the new medium of television had killed many movie theaters. For awhile, this actually helped the maintenance of national cultural policies, because television, in contrast to film, could be controlled through monopoly public broadcast institutions. But this restrictive public system broke down in the 1980s, and the European airwaves and cableways soon filled with even more of Hollywood. Of course, there is also more domestic production of TV programs in each country, as would be expected in a system where there are more channels to fill, where audiences must be won nightly, and where domestic themes and actors attract viewers. Some program ideas were also being copied in the opposite direction. But, on the whole, commercial TV is much more American in content than public TV, whether as actual imports or by being inspired in style.

And now, television over the Internet is knocking. Companies such as Yahoo BB in Japan are in 2003 on the verge of offering affordable video service at decent picture quality to its millions of broadband Internet customers. The question is what will enter when the door is opened? Will it be a multicultural richness of many sources or will it be more of Hollywood?

The knee-jerk response to this question is to invoke Internet platitudes: Anybody can enter, no one can tell a dog on the Internet, a bit is a bit, silicon economics are different than carbon economics, the Internet penetration is higher in Finland than in the United States, and so on. It is as if the Internet community, staunchly internationalist and multicultural by outlook and background, does not want to face the very question of whether it contributes to the further ascendancy of American mass culture. So, what is the answer to that question? It is not an easy one to provide, because it requires an analysis of the future delivery technology, distribution industry, market structure, content formats, and other applications. The real question is, what type of TV will run over the Internet?

## TECHNOLOGY DRIVES THE STRUCTURE OF NETWORKS

For electronic media, transmission technology is destiny: It defines format, content, and economics. It used to be expensive to move information, but now it is cheap. It is possible to do old things in new ways, new things in old ways, and new things in new ways.

### Moving Many Bits

Past technology enabled the creation of transmission networks of two types. The first type moved a lot of bits (strictly speaking, they were analog waveforms), shared by many. Think of it as a fat party line. This is called broadcasting and cable TV, both "synchronous" forms of communication. The second type of network moved a relatively small number of bits, but it did so on an individualized, nonshared, "asynchronous" basis, giving everyone a skinny but individual line. This is called telephony. The two different applications were based on the cost of delivery. People recognized almost from the beginning of TV the usefulness of individualized video transmission. If money were no object, then one could have transmitted individualized video over several phone lines as early as the 1940s. But it was just too expensive to do so outside the labs.

### Moving Bits Over Distance

It also used to be expensive to transmit information across distance, which led to an essentially local form of bit distribution. Broadcast and cable TV were done that way, until satellites came along and enabled regionwide transmission. In telephony, long distance transmission used to be expensive, especially on international routes. But this changed with technology.

The technical elements that brought down the cost of long distance transmission were optical fibers, laser and LED light sources, packet switching, compression algorithms, and microcomponents such as processors and storage devices. On the policy side, market opening and competition were drivers of cost reduction. As a result, cost reduction has been so great that it is often neither metered by distance nor time, and instead offered on a flat-rate basis.

The cheapness of transmission enabled the transmission of text at a price close to zero. This made the narrowband Internet affordable. It moved a relatively limited number of bits at great distances at a low cost. So successful were the applications of the Internet that they created an insatiable hunger for more of bit transport, and in consequence the individualized pipes started to become less skinny.

So now we are in the midst of a historical move: from the *kilobit* stages of individualized communications to that of the *megabit* stage, and within the reasonable future, to the *gigabit* stage. The implications of this transition are as great as the change from a transportation system of railroads to one of automobiles in the 20th century.

## NETWORK STRUCTURE DRIVES THE ECONOMICS OF CONTENT PRODUCTION

These developments have an impact on content. Media content is the kind of information bits for which a sizable number of people would pay in money or attention. It includes live performances, films, TV programs, recorded music, and print publications. It excludes personal correspondence, business documents, baby pictures, home movies, and so forth. Media bit strings are expensive to produce because in order to make them reasonably attractive to audiences, they must be carefully designed, created, and edited.

We will analyze the relative cost of audiovisual media. Each form of delivery has its specific cost characteristics. The calculations are order of magnitude only. There are the costs of the creation of content, which are fixed in nature and largely independent of the actual usage. Then there are the costs of distribution, which usually vary according to the number of users, although they also have a fixed cost component. The discussion begins with live performances.

### Theater

For centuries, audiovisual content was mostly produced through live performances based on edited scripts and scores. Live performance is the yardstick against which the technical performances of all other media are compared.

Consider the cost of producing theater; the latter is defined as a decent regional theater. The cost of producing such content, up to the first curtain call, is about $70 per second of content.[1] This is not trivial, but it is still quite low in comparison to other media, as becomes evident. The real cost problem for theater is its cost of distribution. This distribution cost is .46 cents per viewer and second,[2] or 460 "millicents" per second of transmitting theater content to one viewer. That viewer's incidental costs, such as personal transportation, are not included.

With this distribution cost and its further sensitivity to distance, the reach of each theatrical production is limited, which means that it is possible to establish theater or live performance best where population densities are high or where many people visit, such as New York, London, or Edinburgh. These distribution characteristics make theater a naturally local medium in terms of distribution.

## Film

The high per capita distribution cost of theater led to the film medium. Using today's figures (as for all other media discussed here), production costs are about $50 million per Hollywood film,[3] or about $9,260/sec (a European film costs only about one fifth and an Indian film only one fiftieth of that amount). To distribute the film bits by way of movie theaters, including wholesale distribution and exhibition expenses, comes to a distribution cost of 52 millicents per viewer and per second.

Film, in its Hollywood variety, is thus 130 times as expensive to produce as live quality theater. But it is almost 10 times cheaper to distribute. The cheaper distribution makes it possible to reach more potential viewers and thus amortize the content production cost over a much larger number of people, for a lower total cost. Film is a high fixed-cost, low incremental cost medium relative to theater, and much less distance sensitive. Thus, film is a naturally global medium in economic terms of distribution.

Furthermore, a film production can be distributed in an elaborate sequence of release through various media, such as video rental, pay cable, and TV. A film's content gets circulated first to high paying price inelastic film theater viewers, then down the demand elasticity chain to distribution by other media. The aim is price differentiation among viewers with different elasticities of demand with respect to price. The result is the squeezing

---

[1]Based on $500,000 reproduction, per information for Macarter Theater. Princeton, NJ, by communication.

[2]Calculated from information listed in footnote 1.

[3]Vogel, Harold, *Entertainment Economics,* Cambridge University Press, 2001, Cambridge and New York.

out of a major part of what economists call "consumer surplus." Theater, too, can engage in a release sequence, although such sequence is typically in the *opposite* direction: from off-off Broadway and its equivalents, then off-Broadway and regional theater, and then Broadway and other high-end venues. This means that theater cannot squeeze out most consumer surplus. One reason for this strategy is that there is less willingness to finance the risk of a theatrical production that starts at the top of the distribution chain, because the potential rewards are lower on the upside.

What may be observed is the "death spiral" of small films. People's time and attention is limited, and they allocate it, around the world, to attractive productions that are expensive in terms of production, stars, and so on. As they flock to big-budget films, the average cost of such productions *per viewer* can actually be lower than those of small films. On top of that, the big budget films are better able to be distributed across platforms, and hence appropriate more of the consumer surplus.

European film content is produced at one fifth the cost, but is viewed by one tenth of viewers. And its distribution costs are similar per person. European breakeven points are lower. However, with incremental costs similar or lower, the larger number of potential viewers around the world give American films a much greater upside potential, and this attracts risk capital.

### Broadcast Television

Broadcasting reduces distribution costs dramatically. TV station and network distribution cost per viewer and second is .07 millicent per viewer per second of distribution.[4] This is 750 times cheaper in distribution than film, and 7,000 times cheaper than theater (see Table 16.1). The cost of TV content is about $555/second.[5]

### Cable TV

Distribution costs 0.046 of a millicent per person per second per video channel.[6] This is cheaper than broadcasting on a per-channel basis, because the bundling of numerous channels is cheaper than a station-by-station broadcast distribution, even ignoring the opportunity cost of the spectrum. Cable content is cheaper as well. Content costs for cable-originated programs are $55/second.[7] The cable TV industry is more international in nature than broadcast TV, partly for regulatory reasons (licenses)

---

[4]Based on a $4 million operating budget per station, and a $500 million national network distribution cost.

[5]Based on a $1 million/half-hour network programming, including reruns.

[6]Based on operating costs of $25 per cable household, 50 channels.

[7]Based on $100,000 per half-hour cable channel programming, including reruns.

TABLE 16.1
*Cost*

| | Content/sec ($) | Distrib/Cap/Sec (m¢) |
|---|---|---|
| Theater | 70 | 460 |
| Film | 9,260 | 52 |
| Broadcast | 555 | 0.068 |
| Cable | 110 | 0.046 |
| Internet TV | 110 | 1.85 |

and partly due to economics and technology. Global distribution costs are low and distance insensitive through satellites that function as wholesale distributors. Hence, cable program channels are more global than broadcast channel programs, and many of them are U.S. channels (e.g., MTV, ESPN, CNN, Discovery) with sound tracks in various languages.

### Internet TV

The cost of Internet TV content is hard to estimate. On the one hand, there will be significant need to keep costs down, especially in the early stages. At the same time, the interactivity and multimedia aspect of the medium require additional features beyond straight video. Competition will be fierce for audience share, and commercial providers of Internet TV will have to offer premium level content. Therefore, broadband Internet, available to most households and considered nightly among entertainment options, cannot possibly be produced cheaply.

Hence, program cost of content that is not merely the replay of traditional video will not be lower than that of cable TV, and more likely will be higher. Distribution costs are 1.85 millicent per second and user.[8] This is 40 times higher than the distribution cost per cable channel. The reason is that individualization requires significantly larger transmission resources. A similar disadvantage exists with respect to broadcast TV, where the ratio is 1:27. (Various caching schemes can reduce that ration but at the expense of content diversity.) The implication is that Internet TV can function

---

[8]Based on $40/mo for 1Mbps Internet channel.

economically only as a premium medium, supplemented by premium prices. Several types of applications therefore seem most likely.

1. Internet TV for video-on-demand (VOD) delivery of films, at the very top of the distribution chain, right after movie theater distribution and maybe even ahead. Internet TV is cheaper in distribution than film, which suggests a role for the home as a premium video-on-demand service.

2. Interactivity and multimedia applications (i.e., using the medium in ways that cannot be done over regular, one-way TV).

3. Programs for thin and specialized audiences that would not be served by synchronous TV, and which are willing to pay.

4. Programs subsidized by public sources.

5. Programs supported by commercial sources due to special effectiveness as a marketing medium beyond synchronous TV.

## BASIC ECONOMICS DRIVES APPLICATIONS

This analysis indicates that the cost advantages of cable-style distribution are significant by a factor of about 40. The reduction in distribution cost due to the increasing efficiency of fiber therefore does not mean that all pipes will become individualized. The relative cost of shared (synchronous) transmission is still much lower than that of asynchronous one. Thus, the two will coexist, with the individualized Internet channels providing the premium offerings. What the drop in distance means is that the impact of distance is much reduced and both synchronous and asynchronous networks can be architected for national and global, rather than local, distribution.

From the numbers, it is quite clear that Internet TV should not be used for regular video content distribution. For that purpose, cable TV and its digital fiber variants will be much cheaper, especially in combination with a personal video recorder (PVR). Internet TV's market is for applications that go beyond regular TV: interactivity, asynchronicity, linkages, multimedia, or communications. To produce such content is expensive. It requires creativity, lots of programmers, significant alpha and beta testing, and continuous innovation. Such high-cost content exhibits strong economies of scale on the content production side, and network externalities on the demand side. Both favor providers that can come up with big budgets, can diversify risk, distribute also over other platforms, create tie-ins, and establish user communities. Even for nonpremium programs (i.e., creative small productions) or sex shows and games, where the absolute production costs are lower, the advantages of a large user base still apply.

The United States has a large Internet community with entrepreneurial energy, big content producing companies with worldwide distribution and experience in reaching popular audiences, creative and technological talent from all over the world, and efficient production clusters; it also benefits from the *lingua franca* advantages of the English language and the cultural prowess of being the world's superpower. There are also leading computer hardware, components, and telecom industries, a pro-competition push, and a financial system that provides risk capital. Some of these factors are also available elsewhere, but nowhere in such combination.

Thus, the medium of Internet TV combines the strengths of the U.S. economy and society in entertainment content, in Internet, and in e-transactions. Add to that economies of scale, and there is nothing on the horizon that can match it. And, therefore, Internet TV will be strongly American. Participants from other countries will also be players, but most likely either domestically without much reach, or global players who will offer basically American-style content to the world, like sitcoms and the Italian "spaghetti westerns" of the past. (Of course, the pipes are not one-way streets, and they could be used for content produced elsewhere to be distributed globally, and into the United States. But to do so, the provider would have modify domestic content to create a global attractiveness, and evolve into "mid-Atlantic," "mid-Pacific" style of content.

Thus, there will be winners and losers. The losers will not sit still, but they will invoke various public policy concerns, which will inevitably lead to protectionism. Therefore, it is necessary to be ready for cultural and trade wars of the Internet TV of the near future.

# Author Index

243

# Subject Index

**247**

T - #0506 - 101024 - C0 - 229/152/15 - PB - 9780805843064 - Gloss Lamination